ArcGIS® 9

Understanding Map Projections

Contents

1 Geographic coordinate systems

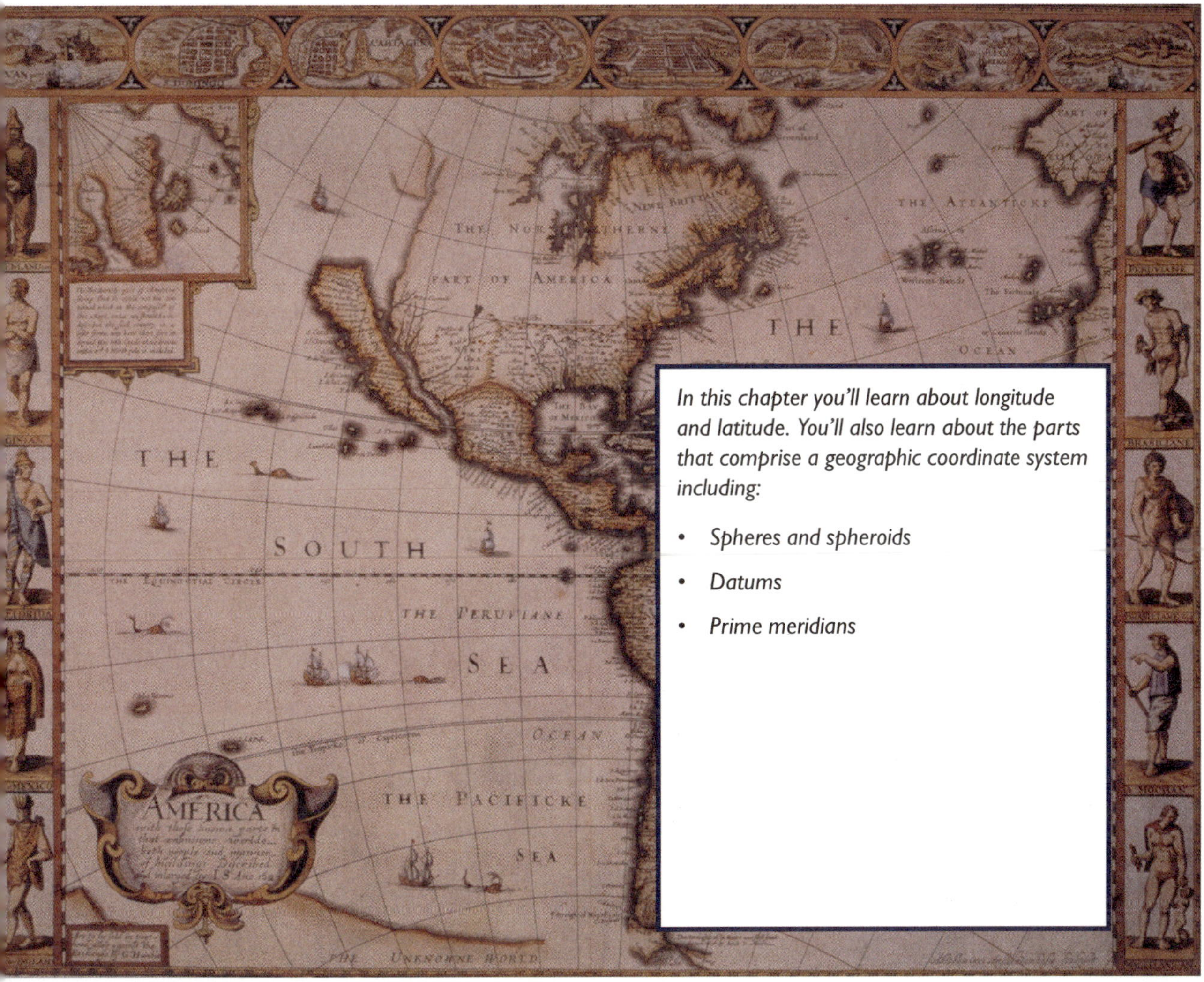

In this chapter you'll learn about longitude and latitude. You'll also learn about the parts that comprise a geographic coordinate system including:

- *Spheres and spheroids*
- *Datums*
- *Prime meridians*

A *geographic coordinate system* (GCS) uses a three-dimensional spherical surface to define locations on the earth. A GCS is often incorrectly called a datum, but a datum is only one part of a GCS. A GCS includes an angular unit of measure, a prime meridian, and a datum (based on a spheroid).

A point is referenced by its *longitude* and *latitude* values. Longitude and latitude are angles measured from the earth's center to a point on the earth's surface. The angles often are measured in degrees (or in grads).

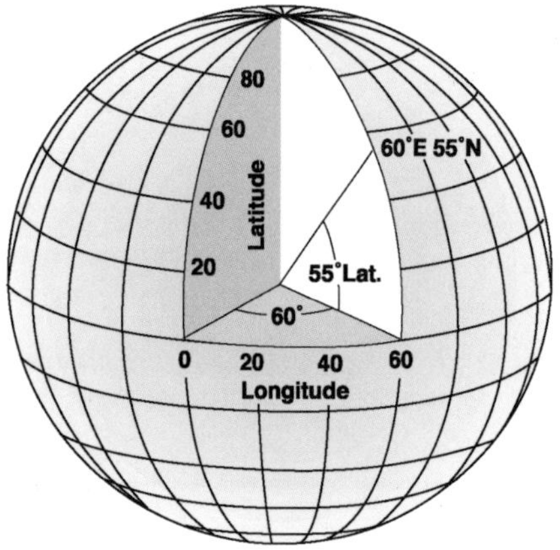

The world as a globe showing the longitude and latitude values.

In the spherical system, horizontal lines, or east–west lines, are lines of equal latitude, or *parallels*. Vertical lines, or north–south lines, are lines of equal longitude, or *meridians*. These lines encompass the globe and form a gridded network called a *graticule*.

The line of latitude midway between the poles is called the equator. It defines the line of zero latitude. The line of zero longitude is called the prime meridian. For most geographic coordinate systems, the prime meridian is the longitude that passes through Greenwich, England. Other countries use longitude lines that pass through Bern, Bogota, and Paris as prime meridians.

The origin of the graticule (0,0) is defined by where the equator and prime meridian intersect. The globe is then divided into four geographical quadrants that are based on compass bearings from the origin. North and south are above and below the equator, and west and east are to the left and right of the prime meridian.

Latitude and longitude values are traditionally measured either in decimal degrees or in degrees, minutes, and seconds (DMS). Latitude values are measured relative to the equator and range from -90° at the South Pole to +90° at the North Pole. Longitude values are measured relative to the prime meridian. They range from -180° when traveling west to 180° when traveling east. If the prime meridian is at Greenwich, then Australia, which is south of the equator and east of Greenwich, has positive longitude values and negative latitude values.

Although longitude and latitude can locate exact positions on the surface of the globe, they are not uniform units of measure. Only along the equator does the distance represented by one degree of longitude

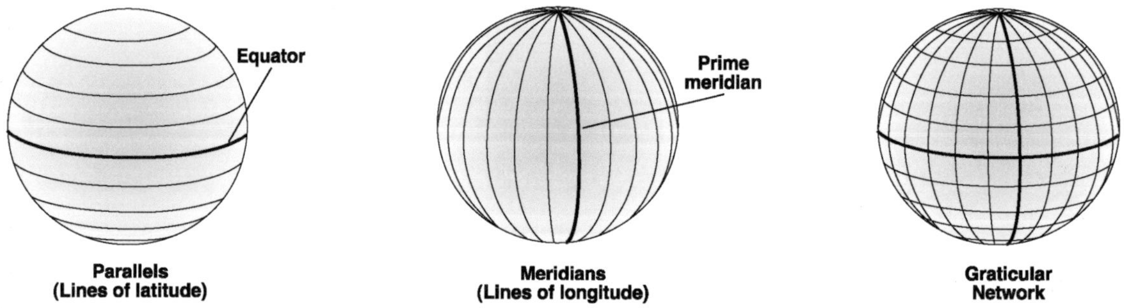

The parallels and meridians that form a graticule.

approximate the distance represented by one degree of latitude. This is because the equator is the only parallel as large as a meridian. (Circles with the same radius as the spherical earth are called *great circles*. The equator and all meridians are great circles.)

Above and below the equator, the circles defining the parallels of latitude get gradually smaller until they become a single point at the North and South Poles where the meridians converge. As the meridians converge toward the poles, the distance represented by one degree of longitude decreases to zero. On the Clarke 1866 spheroid, one degree of longitude at the equator equals 111.321 km, while at 60° latitude it is only 55.802 km. Since degrees of latitude and longitude don't have a standard length, you can't measure distances or areas accurately or display the data easily on a flat map or computer screen.

The shape and size of a geographic coordinate system's surface is defined by a sphere or spheroid. Although the earth is best represented by a spheroid, the earth is sometimes treated as a sphere to make mathematical calculations easier. The assumption that the earth is a sphere is possible for small-scale maps (smaller than 1:5,000,000). At this scale, the difference between a sphere and a spheroid is not detectable on a map. However, to maintain accuracy for larger-scale maps (scales of 1:1,000,000 or larger), a spheroid is necessary to represent the shape of the earth. Between those scales, choosing to use a sphere or spheroid will depend on the map's purpose and the accuracy of the data.

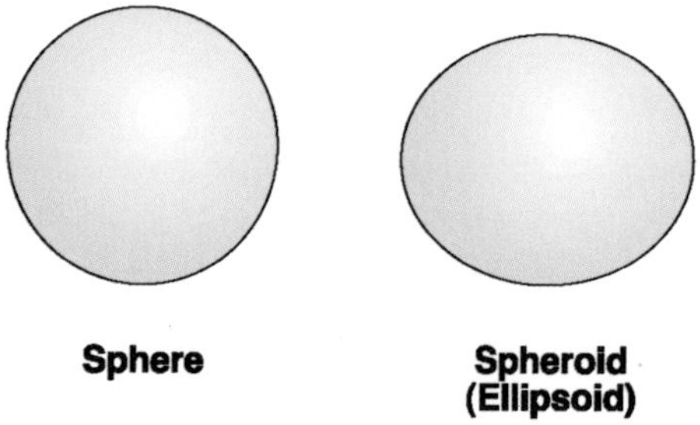

A sphere is based on a circle, while a spheroid (or ellipsoid) is based on an ellipse. The shape of an ellipse is defined by two radii. The longer radius is called the semimajor axis, and the shorter radius is called the semiminor axis.

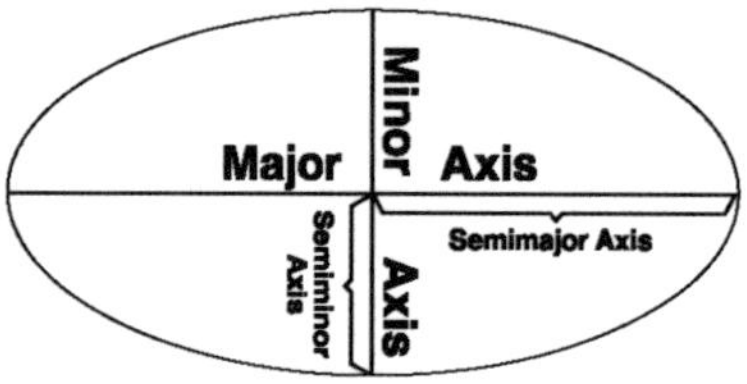

The major and minor axes of an ellipse.

Rotating the ellipse around the semiminor axis creates a spheroid. A spheroid is also known as an oblate ellipsoid of revolution.

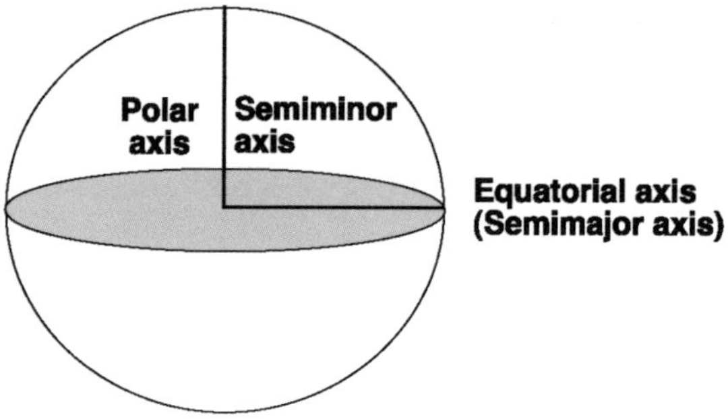

The semimajor axis and semiminor axis of a spheroid.

A spheroid is defined by either the semimajor axis, *a,* and the semiminor axis, *b,* or by *a* and the *flattening*. The flattening is the difference in length between the two axes expressed as a fraction or a decimal. The flattening, *f,* is:

```
f = (a - b) / a
```

The flattening is a small value, so usually the quantity *1/f* is used instead. The spheroid parameters for the World Geodetic System of 1984 (WGS 1984 or WGS84) are:

```
a   =  6378137.0  meters
1/f  =  298.257223563
```

The flattening ranges from zero to one. A flattening value of zero means the two axes are equal, resulting in a sphere. The flattening of the earth is approximately 0.003353.

Another quantity, that, like the flattening, describes the shape of a spheroid, is the square of the *eccentricity*, e^2. It is represented by:

$$e^2 = \frac{a^2 - b^2}{a^2}$$

DEFINING DIFFERENT SPHEROIDS FOR ACCURATE MAPPING

The earth has been surveyed many times to help us better understand its surface features and their peculiar irregularities. The surveys have resulted in many spheroids that represent the earth. Generally, a spheroid is chosen to fit one country or a particular area. A spheroid that best fits one region is not necessarily the same one that fits another region. Until recently, North American data used a spheroid determined by Clarke in 1866. The semimajor axis of the Clarke 1866 spheroid is 6,378,206.4 meters, and the semiminor axis is 6,356,583.8 meters.

Because of gravitational and surface feature variations, the earth is neither a perfect sphere nor a perfect spheroid. Satellite technology has revealed several elliptical deviations; for example, the South Pole is closer to the equator than the North Pole. Satellite-determined spheroids are replacing the older ground-measured spheroids. For example, the new standard spheroid for North America is the Geodetic Reference System of 1980 (GRS 1980), whose radii are 6,378,137.0 and 6,356,752.31414 meters.

Because changing a coordinate system's spheroid changes all previously measured values, many organizations haven't switched to newer (and more accurate) spheroids.

While a spheroid approximates the shape of the earth, a datum defines the position of the spheroid relative to the center of the earth. A datum provides a frame of reference for measuring locations on the surface of the earth. It defines the origin and orientation of latitude and longitude lines.

Whenever you change the datum, or more correctly, the geographic coordinate system, the coordinate values of your data will change. Here's the coordinates in DMS of a control point in Redlands, California, on the North American Datum of 1983 (NAD 1983 or NAD83).

```
-117  12  57.75961    34  01  43.77884
```

Here's the same point on the North American Datum of 1927 (NAD 1927 or NAD27).

```
-117  12  54.61539    34  01  43.72995
```

The longitude value differs by about three seconds, while the latitude value differs by about 0.05 seconds.

In the last 15 years, satellite data has provided geodesists with new measurements to define the best earth-fitting spheroid, which relates coordinates to the earth's center of mass. An earth-centered, or geocentric, datum uses the earth's center of mass as the origin. The most recently developed and widely used datum is WGS 1984. It serves as the framework for locational measurement worldwide.

A local datum aligns its spheroid to closely fit the earth's surface in a particular area. A point on the surface of the spheroid is matched to a particular position on the surface of the earth. This point is known as the origin point of the datum. The coordinates of the origin point are fixed, and all other points are calculated from it. The coordinate system origin of a local datum is not at the center of the earth. The center of the spheroid of a local datum is offset from the earth's center. NAD 1927 and the European Datum of 1950 (ED 1950) are local datums. NAD 1927 is designed to fit North America reasonably well, while ED 1950 was created for use in Europe. Because a local datum aligns its spheroid so closely to a particular area on the earth's surface, it's not suitable for use outside the area for which it was designed.

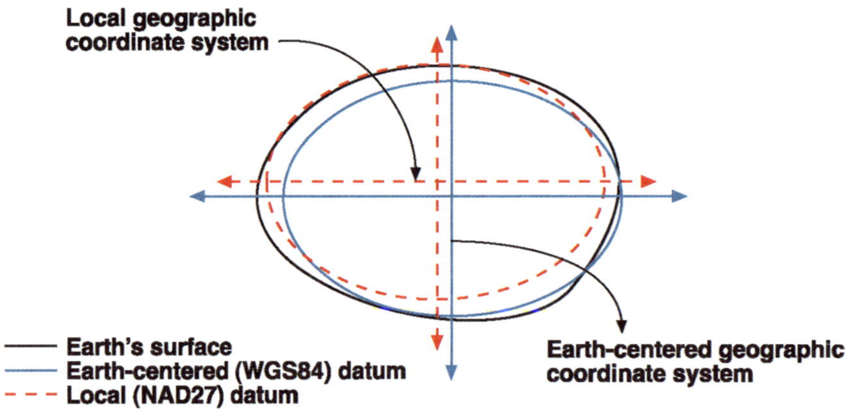

The two horizontal datums used almost exclusively in North America are NAD 1927 and NAD 1983.

NAD 1927

NAD 1927 uses the Clarke 1866 spheroid to represent the shape of the earth. The origin of this datum is a point on the earth referred to as Meades Ranch in Kansas. Many NAD 1927 control points were calculated from observations taken in the 1800s. These calculations were done manually and in sections over many years. Therefore, errors varied from station to station.

NAD 1983

Many technological advances in surveying and geodesy—electronic theodolites, Global Positioning System (GPS) satellites, Very Long Baseline Interferometry, and Doppler systems—revealed weaknesses in the existing network of control points. Differences became particularly noticeable when linking existing control with newly established surveys. The establishment of a new datum allowed a single datum to cover consistently North America and surrounding areas.

The North American Datum of 1983 is based on both earth and satellite observations, using the GRS 1980 spheroid. The origin for this datum is the earth's center of mass. This affects the surface location of all longitude–latitude values enough to cause locations of previous control points in North America to shift, sometimes as much as 500 feet. A 10-year multinational effort tied together a network of control points for the United States, Canada, Mexico, Greenland, Central America, and the Caribbean.

The GRS 1980 spheroid is almost identical to the WGS 1984 spheroid. The WGS 1984 and NAD 1983 coordinate systems are both earth-centered. Because both are so close, NAD 1983 is compatible with GPS data. The raw GPS data is actually reported in the WGS 1984 coordinate system.

HARN OR HPGN

There is an ongoing effort at the state level to readjust the NAD 1983 datum to a higher level of accuracy using state-of-the-art surveying techniques that were not widely available when the NAD 1983 datum was being developed. This effort, known as the High Accuracy Reference Network (HARN), or High Precision Geodetic Network (HPGN), is a cooperative project between the National Geodetic Survey and the individual states.

Currently, all states have been resurveyed, but not all of the data has been released to the public. As of September 2000, the grids for 44 states and two territories have been published.

OTHER UNITED STATES DATUMS

Alaska, Hawaii, Puerto Rico and the Virgin Islands, and some Alaskan islands have used other datums besides NAD 1927. See Chapter 3, 'Geographic transformations', for more information. New data is referenced to NAD 1983.

2 Projected coordinate systems

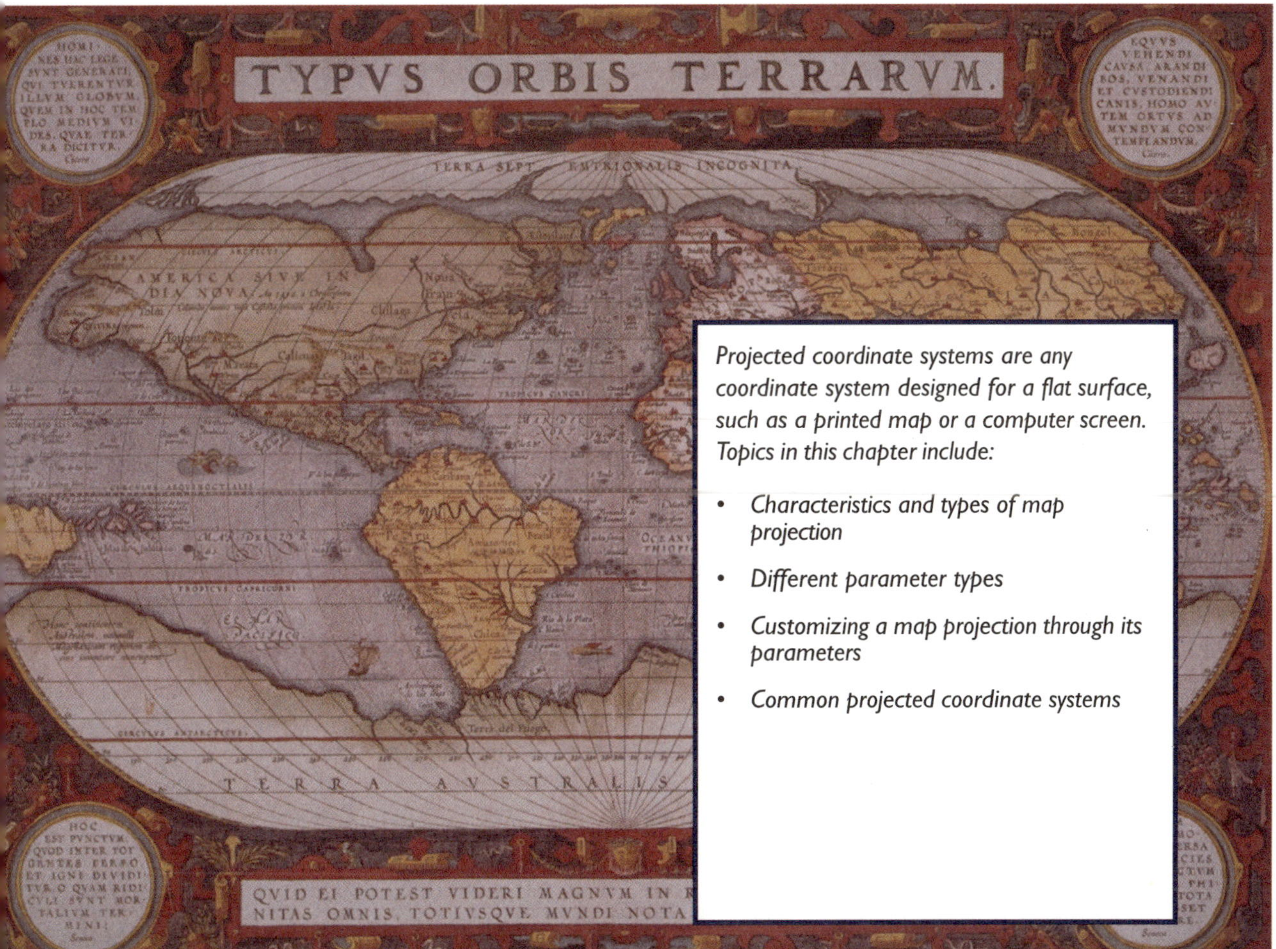

Projected coordinate systems are any coordinate system designed for a flat surface, such as a printed map or a computer screen. Topics in this chapter include:

- *Characteristics and types of map projection*
- *Different parameter types*
- *Customizing a map projection through its parameters*
- *Common projected coordinate systems*

A projected coordinate system is defined on a flat, two-dimensional surface. Unlike a geographic coordinate system, a projected coordinate system has constant lengths, angles, and areas across the two dimensions. A projected coordinate system is always based on a geographic coordinate system that is based on a sphere or spheroid.

In a projected coordinate system, locations are identified by x,y coordinates on a grid, with the origin at the center of the grid. Each position has two values that reference it to that central location. One specifies its horizontal position and the other its vertical position. The two values are called the x-coordinate and y-coordinate. Using this notation, the coordinates at the origin are x = 0 and y = 0.

On a gridded network of equally spaced horizontal and vertical lines, the horizontal line in the center is called the x-axis and the central vertical line is called the y-axis. Units are consistent and equally spaced across the full range of x and y. Horizontal lines above the origin and vertical lines to the right of the origin have positive values; those below or to the left have negative values. The four quadrants represent the four possible combinations of positive and negative x- and y-coordinates.

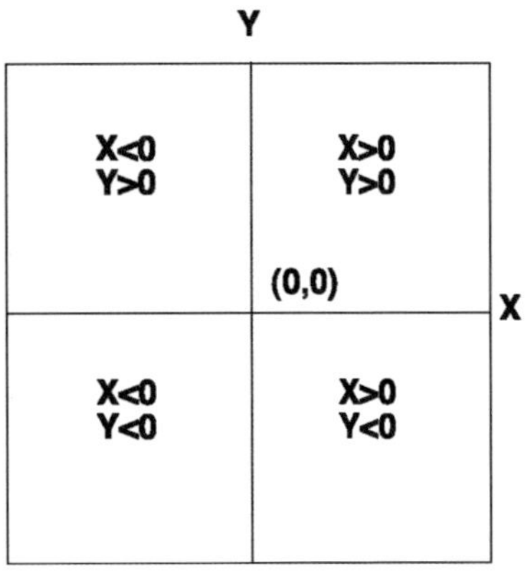

The signs of x,y coordinates in a projected coordinate system.

Whether you treat the earth as a sphere or a spheroid, you must transform its three-dimensional surface to create a flat map sheet. This mathematical transformation is commonly referred to as a *map projection*. One easy way to understand how map projections alter spatial properties is to visualize shining a light through the earth onto a surface, called the projection surface. Imagine the earth's surface is clear with the graticule drawn on it. Wrap a piece of paper around the earth. A light at the center of the earth will cast the shadows of the graticule onto the piece of paper. You can now unwrap the paper and lay it flat. The shape of the graticule on the flat paper is very different than on the earth. The map projection has distorted the graticule.

A spheroid can't be flattened to a plane any easier than a piece of orange peel can be flattened—it will rip.

Representing the earth's surface in two dimensions causes distortion in the shape, area, distance, or direction of the data.

A map projection uses mathematical formulas to relate spherical coordinates on the globe to flat, planar coordinates.

Different projections cause different types of distortions. Some projections are designed to minimize the distortion of one or two of the data's characteristics. A projection could maintain the area of a feature but alter its shape. In the graphic below, data near the poles is stretched. The diagram on the next page shows how three-dimensional features are compressed to fit onto a flat surface.

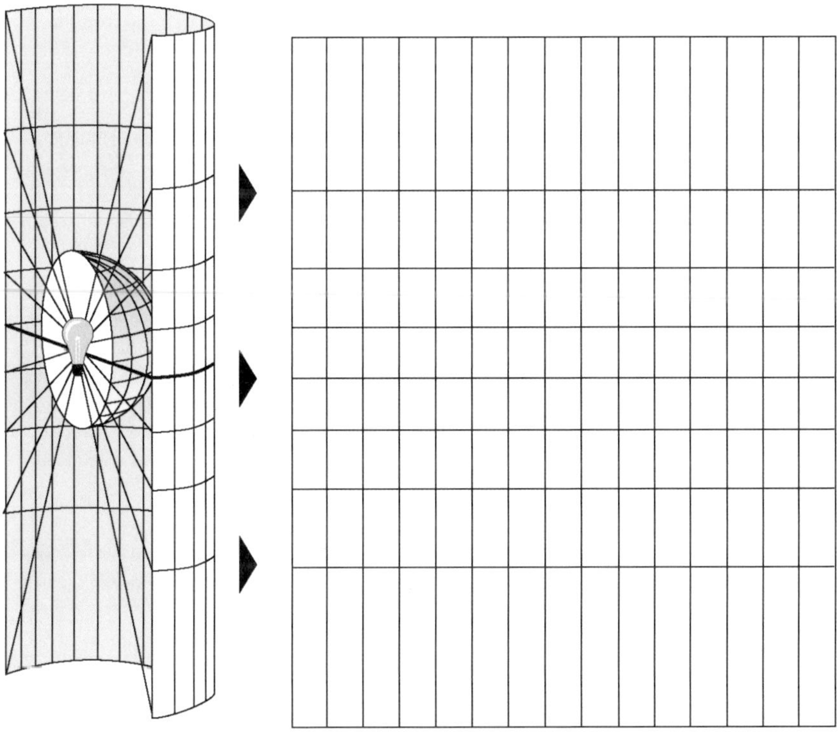

The graticule of a geographic coordinate system is projected onto a cylindrical projection surface.

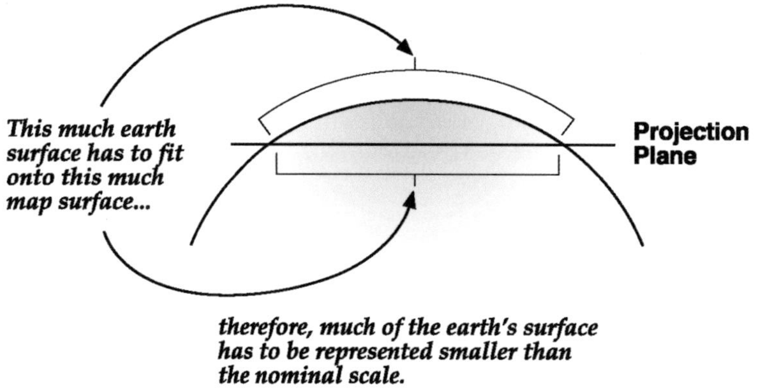

Map projections are designed for specific purposes. One map projection might be used for large-scale data in a limited area, while another is used for a small-scale map of the world. Map projections designed for small-scale data are usually based on spherical rather than spheroidal geographic coordinate systems.

Conformal projections

Conformal projections preserve local shape. To preserve individual angles describing the spatial relationships, a conformal projection must show the perpendicular graticule lines intersecting at 90-degree angles on the map. A map projection accomplishes this by maintaining all angles. The drawback is that the area enclosed by a series of arcs may be greatly distorted in the process. No map projection can preserve shapes of larger regions.

Equal area projections

Equal area projections preserve the area of displayed features. To do this, the other properties—shape, angle, and scale—are distorted. In equal area projections, the meridians and parallels may not intersect at right angles. In some instances, especially maps of smaller regions, shapes are not obviously distorted, and distinguishing an equal area projection from a conformal projection is difficult unless documented or measured.

Equidistant projections

Equidistant maps preserve the distances between certain points. Scale is not maintained correctly by any projection throughout an entire map; however, there are, in most cases, one or more lines on a map along which scale is maintained correctly. Most equidistant projections have one or more lines for which the length of the line on a map is the same length (at map scale) as the same line on the globe, regardless of whether it is a great or small circle or straight or curved. Such distances are said to be *true*. For example, in the Sinusoidal projection, the equator and all parallels are their true lengths. In other equidistant projections, the equator and all meridians are true. Still others (e.g., Two-Point Equidistant) show true scale between one or two points and every other point on the map. Keep in mind that no projection is equidistant to and from all points on a map.

True-direction projections

The shortest route between two points on a curved surface such as the earth is along the spherical equivalent of a straight line on a flat surface. That is the great circle on which the two points lie. True-direction, or *azimuthal*, projections maintain some of the great circle arcs, giving the directions or azimuths of all points on the map correctly with respect to the center. Some true-direction projections are also conformal, equal area, or equidistant.

Because maps are flat, some of the simplest projections are made onto geometric shapes that can be flattened without stretching their surfaces. These are called developable surfaces. Some common examples are cones, cylinders, and planes. A map projection systematically projects locations from the surface of a spheroid to representative positions on a flat surface using mathematical algorithms.

The first step in projecting from one surface to another is creating one or more points of contact. Each contact is called a point (or line) of tangency. As illustrated in the section about 'Planar projections' below, a planar projection is tangential to the globe at one point. Tangential cones and cylinders touch the globe along a line. If the projection surface intersects the globe instead of merely touching its surface, the resulting projection is a secant rather than a tangent case. Whether the contact is tangent or secant, the contact points or lines are significant because they define locations of zero distortion. Lines of true scale are often referred to as *standard lines*. In general, distortion increases with the distance from the point of contact.

Many common map projections are classified according to the projection surface used: conic, cylindrical, or planar.

Conic projections

The most simple conic projection is tangent to the globe along a line of latitude. This line is called the *standard parallel*. The meridians are projected onto the conical surface, meeting at the apex, or point, of the cone. Parallel lines of latitude are projected onto the cone as rings. The cone is then cut along any meridian to produce the final conic projection, which has straight converging lines for meridians and concentric circular arcs for parallels. The meridian opposite the cut line becomes the *central meridian*.

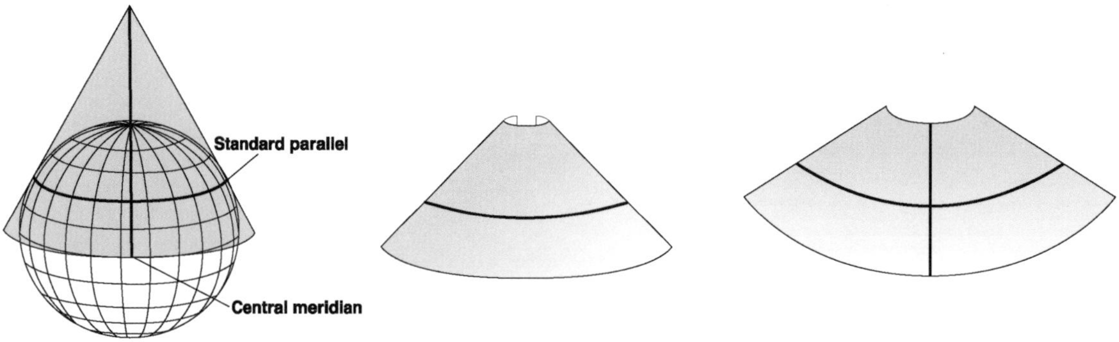

In general, the further you get from the standard parallel, the more distortion increases. Thus, cutting off the top of the cone produces a more accurate projection. You can accomplish this by not using the polar region of the projected data. Conic projections are used for midlatitude zones that have an east–west orientation.

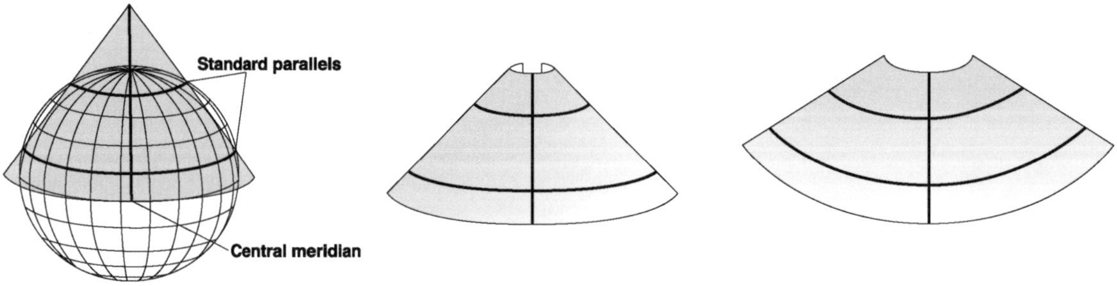

Somewhat more complex conic projections contact the global surface at two locations. These projections are called *secant projections* and are defined by two standard parallels. It is also possible to define a secant projection by one standard parallel and a scale factor. The distortion pattern for secant projections is different between the standard parallels than beyond them. Generally, a secant projection has less overall distortion than a tangent projection. On still more complex conic projections, the axis of the cone does not line up with the polar axis of the globe. These types of projections are called *oblique*.

The representation of geographic features depends on the spacing of the parallels. When equally spaced, the projection is equidistant north–south but neither conformal nor equal area. An example of this type of projection is the Equidistant Conic projection. For small areas, the overall distortion is minimal. On the Lambert Conic Conformal projection, the central parallels are spaced more closely than the parallels near the border, and small geographic shapes are maintained for both small-scale and large-scale maps. On the Albers Equal Area

Conic projection, the parallels near the northern and southern edges are closer together than the central parallels, and the projection displays equivalent areas.

Cylindrical projections

Like conic projections, cylindrical projections can also have tangent or secant cases. The Mercator projection is one of the most common cylindrical projections, and the equator is usually its line of tangency. Meridians are geometrically projected onto the cylindrical surface, and parallels are mathematically projected. This produces graticular angles of 90 degrees. The cylinder is cut along any meridian to produce the final cylindrical projection. The meridians are equally spaced, while the spacing between parallel lines of latitude increases toward the poles. This projection is conformal and displays true direction along straight lines. On a Mercator projection, *rhumb lines*, lines of constant bearing, are straight lines, but most great circles are not.

For more complex cylindrical projections the cylinder is rotated, thus changing the tangent or secant lines. Transverse cylindrical projections such as the Transverse Mercator use a meridian as the tangential contact or lines parallel to meridians as lines of secancy. The standard lines then run north–south, along which the scale is true. Oblique cylinders are rotated around a great circle line located anywhere between the equator and the meridians. In these more complex projections, most meridians and lines of latitude are no longer straight.

In all cylindrical projections, the line of tangency or lines of secancy have no distortion and thus are lines of equidistance. Other geographical properties vary according to the specific projection.

Normal

Transverse

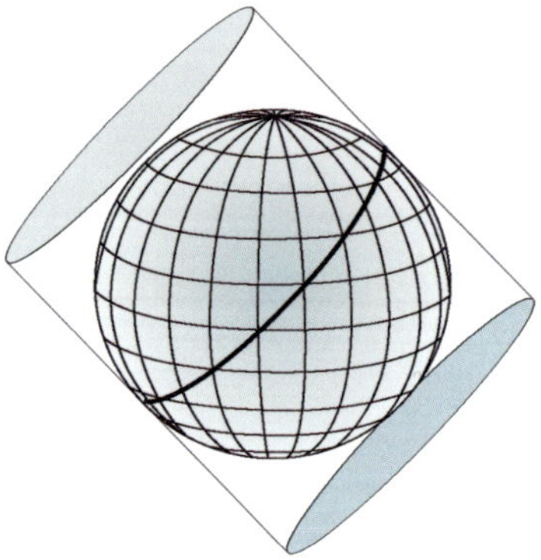

Oblique

Planar projections

Planar projections project map data onto a flat surface touching the globe. A planar projection is also known as an azimuthal projection or a zenithal projection. This type of projection is usually tangent to the globe at one point but may be secant, also. The point of contact may be the North Pole, the South Pole, a point on the equator, or any point in between. This point specifies the aspect and is the focus of the projection. The focus is identified by a central longitude and a central latitude. Possible aspects are *polar*, *equatorial*, and *oblique*.

Azimuthal projections are classified in part by the focus and, if applicable, by the perspective point. The graphic below compares three planar projections with polar aspects but different perspectives. The Gnomonic projection views the surface data from the center of the

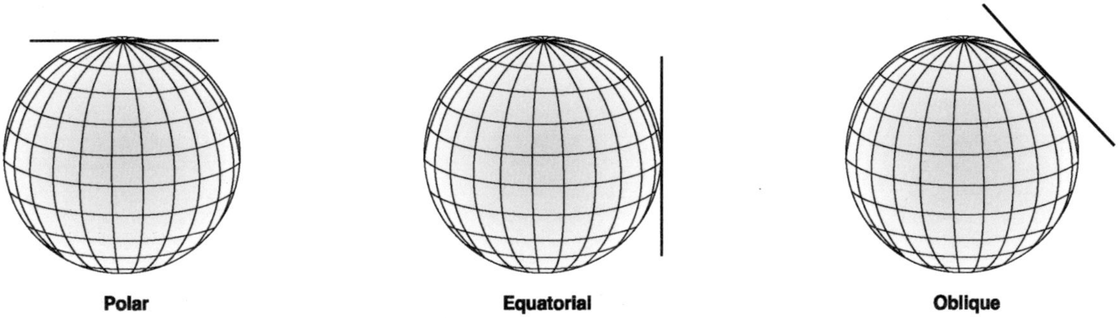

Polar aspects are the simplest form. Parallels of latitude are concentric circles centered on the pole, and meridians are straight lines that intersect with their true angles of orientation at the pole. In other aspects, planar projections will have graticular angles of 90 degrees at the focus. Directions from the focus are accurate.

Great circles passing through the focus are represented by straight lines; thus the shortest distance from the center to any other point on the map is a straight line. Patterns of area and shape distortion are circular about the focus. For this reason, azimuthal projections accommodate circular regions better than rectangular regions. Planar projections are used most often to map polar regions.

Some planar projections view surface data from a specific point in space. The point of view determines how the spherical data is projected onto the flat surface. The perspective from which all locations are viewed varies between the different azimuthal projections. The perspective point may be the center of the earth, a surface point directly opposite from the focus, or a point external to the globe, as if seen from a satellite or another planet.

earth, whereas the Stereographic projection views it from pole to pole. The Orthographic projection views the earth from an infinite point, as if from deep space. Note how the differences in perspective determine the amount of distortion toward the equator.

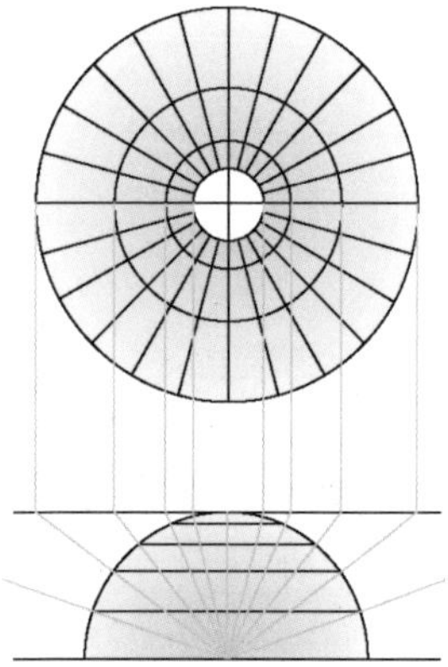

Gnomonic

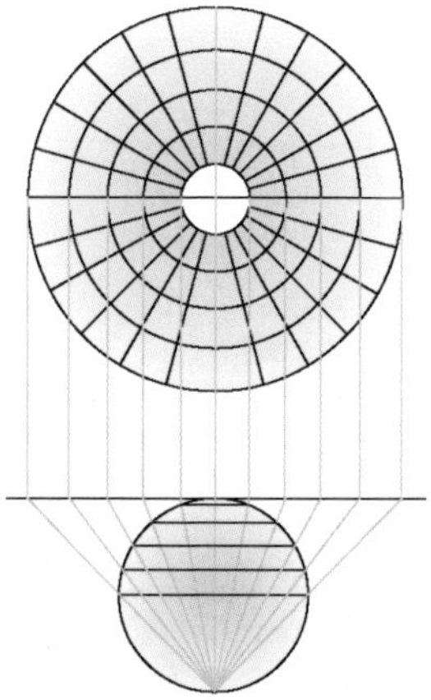

Stereographic

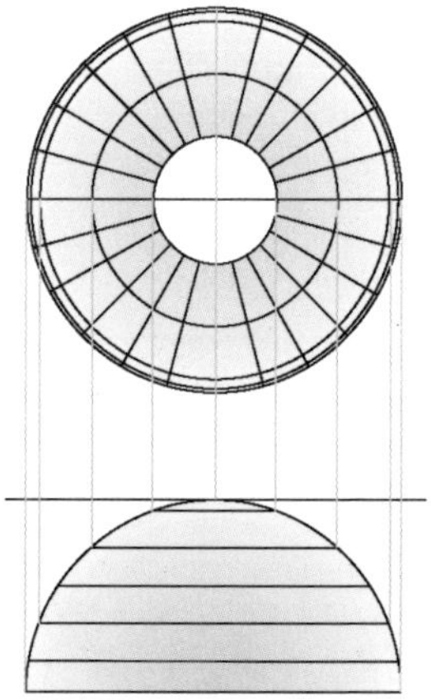

Orthographic

Other projections

The projections discussed previously are conceptually created by projecting from one geometric shape (a sphere) onto another (a cone, cylinder, or plane). Many projections are not related as easily to a cone, cylinder, or plane.

Modified projections are altered versions of other projections (e.g., the Space Oblique Mercator is a modification of the Mercator projection). These modifications are made to reduce distortion, often by including additional standard lines or changing the distortion pattern.

Pseudo projections have some of the characteristics of another class of projection. For example, the Sinusoidal is called a pseudocylindrical projection because all lines of latitude are straight and parallel and all meridians are equally spaced. However, it is not truly a cylindrical projection because all meridians except the central meridian are curved. This results in a map of the earth having an oval shape instead of a rectangular shape.

Other projections are assigned to special groups, such as circular or star.

A map projection by itself isn't enough to define a projected coordinate system. You can state that a dataset is in Transverse Mercator, but that's not enough information. Where is the center of the projection? Was a scale factor used? Without knowing the exact values for the projection parameters, the dataset can't be reprojected.

You can also get some idea of the amount of distortion the projection has added to the data. If you're interested in Australia but you know that a dataset's projection is centered at 0,0, the intersection of the equator and the Greenwich prime meridian, you might want to think about changing the center of the projection.

Each map projection has a set of parameters that you must define. The parameters specify the origin and customize a projection for your area of interest. Angular parameters use the geographic coordinate system units, while linear parameters use the projected coordinate system units.

Linear parameters

False easting—A linear value applied to the origin of the x-coordinates.

False northing—A linear value applied to the origin of the y-coordinates.

False easting and northing values are usually applied to ensure that all x or y values are positive. You can also use the false easting and northing parameters to reduce the range of the x- or y-coordinate values. For example, if you know all y values are greater than five million meters, you could apply a false northing of -5,000,000.

Height—Defines the point of perspective above the surface of the sphere or spheroid for the Vertical Near-side Perspective projection.

Angular parameters

Azimuth—Defines the center line of a projection. The rotation angle measures east from north. Used with the Azimuth cases of the Hotine Oblique Mercator projection.

Central meridian—Defines the origin of the x-coordinates.

Longitude of origin—Defines the origin of the x-coordinates. The central meridian and longitude of origin parameters are synonymous.

Central parallel—Defines the origin of the y-coordinates.

Latitude of origin—Defines the origin of the y-coordinates. This parameter may not be located at the center of the projection. In particular, conic projections use this parameter to set the origin of the y-coordinates below the area of the interest. In that instance, you don't need to set a false northing parameter to ensure that all y-coordinates are positive.

Longitude of center—Used with the Hotine Oblique Mercator Center (both Two-Point and Azimuth) cases to define the origin of the x-coordinates. Usually synonymous with the longitude of origin and central meridian parameters.

Latitude of center—Used with the Hotine Oblique Mercator Center (both Two-Point and Azimuth) cases to define the origin of the y-coordinates. It is almost always the center of the projection.

Standard parallel 1 and standard parallel 2—Used with conic projections to define the latitude lines where the scale is 1.0. When defining a Lambert Conformal Conic projection with one standard parallel, the first standard parallel defines the origin of the y-coordinates.

For other conic cases, the y-coordinate origin is defined by the latitude of origin parameter.

Longitude of first point
Latitude of first point
Longitude of second point
Latitude of second point

The four parameters above are used with the Two-Point Equidistant and Hotine Oblique Mercator projections. They specify two geographic points that define the center axis of a projection.

Pseudo standard parallel 1—Used in the Krovak projection to define the oblique cone's standard parallel.

XY plane rotation—Along with the X scale and Y scale parameters, defines the orientation of the Krovak projection.

Unitless parameters

Scale factor—A unitless value applied to the center point or line of a map projection.

The scale factor is usually slightly less than one. The UTM coordinate system, which uses the Transverse Mercator projection, has a scale factor of 0.9996. Rather than 1.0, the scale along the central meridian of the projection is 0.9996. This creates two almost parallel lines approximately 180 kilometers away, where the scale is 1.0. The scale factor reduces the overall distortion of the projection in the area of interest.

X scale—Used in the Krovak projection to orient the axes.

Y scale—Used in the Krovak projection to orient the axes.

Option—Used in the Cube and Fuller projections. In the Cube projection, option defines the location of the polar facets. An option of 0 in the Fuller projection will display all 20 facets. Specifying an option value between 1–20 will display a single facet.

3 Geographic transformations

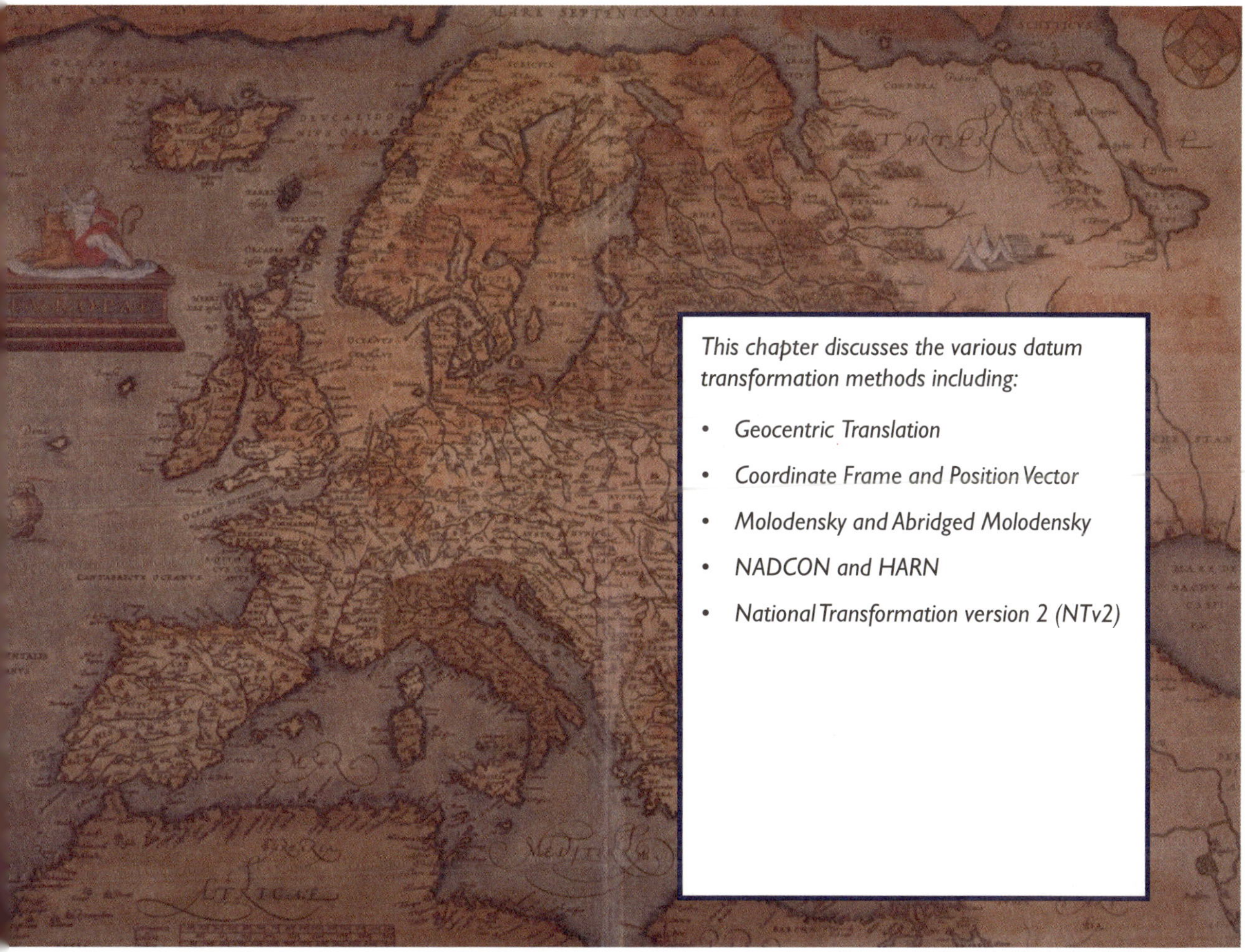

This chapter discusses the various datum transformation methods including:

- *Geocentric Translation*
- *Coordinate Frame and Position Vector*
- *Molodensky and Abridged Molodensky*
- *NADCON and HARN*
- *National Transformation version 2 (NTv2)*

Moving your data between coordinate systems sometimes includes transforming between the geographic coordinate systems.

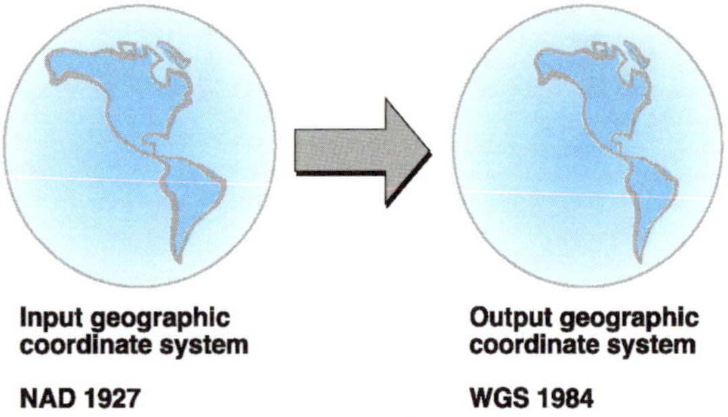

Because the geographic coordinate systems contain datums that are based on spheroids, a geographic transformation also changes the underlying spheroid. There are several methods, which have different levels of accuracy and ranges, for transforming between datums. The accuracy of a particular transformation can range from centimeters to meters depending on the method and the quality and number of control points available to define the transformation parameters.

A geographic transformation always converts geographic (longitude–latitude) coordinates. Some methods convert the geographic coordinates to geocentric (X,Y,Z) coordinates, transform the X,Y,Z coordinates, and convert the new values back to geographic coordinates.

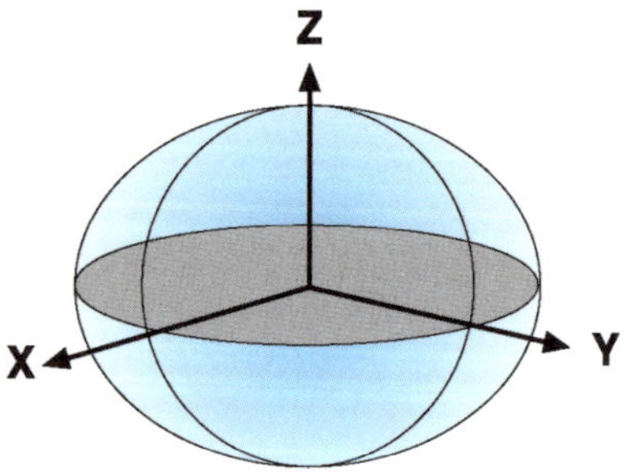

The X,Y,Z coordinate system.

These include the Geocentric Translation, Molodensky, and Coordinate Frame methods.

Other methods such as NADCON and NTv2 use a grid of differences and convert the longitude–latitude values directly.

A geographic transformation is always defined with a direction. The transformation parameters describe how to convert from the input geographic coordinate to the output geographic coordinate system. All supported methods are invertible. Given a geographic transformation, you can apply it in the opposite direction. Generally, applications will automatically apply the transformation in the appropriate direction. As an example, if you wish to convert data from WGS 1984 to Adindan but a list of available geographic transformations shows Adindan_To_WGS_1984, you can choose this transformation and the application will apply it properly.

Three-parameter methods

The simplest datum transformation method is a geocentric, or three-parameter, transformation. The geocentric transformation models the differences between two datums in the X,Y,Z coordinate system. One datum is defined with its center at 0,0,0. The center of the other datum is defined at some distance (ΔX,ΔY,ΔZ) in meters away.

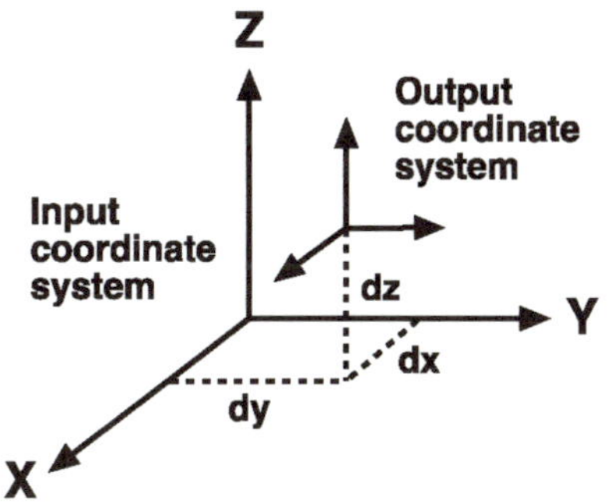

Usually the transformation parameters are defined as going from a local datum to WGS 1984 or another geocentric datum.

$$\begin{bmatrix} X \\ Y \\ Z \end{bmatrix}_{new} = \begin{bmatrix} \Delta X \\ \Delta Y \\ \Delta Z \end{bmatrix} + \begin{bmatrix} X \\ Y \\ Z \end{bmatrix}_{original}$$

The three parameters are linear shifts and are always in meters.

Seven-parameter methods

A more complex and accurate datum transformation is possible by adding four more parameters to a geocentric transformation. The seven parameters are three linear shifts (ΔX,ΔY,ΔZ), three angular rotations around each axis (r_x,r_y,r_z), and scale factor(s).

$$\begin{bmatrix} X \\ Y \\ Z \end{bmatrix}_{new} = \begin{bmatrix} \Delta X \\ \Delta Y \\ \Delta Z \end{bmatrix} + (1+s) \cdot \begin{bmatrix} 1 & r_z & -r_y \\ -r_z & 1 & r_x \\ r_y & -r_x & 1 \end{bmatrix} \cdot \begin{bmatrix} X \\ Y \\ Z \end{bmatrix}_{original}$$

The rotation values are given in decimal seconds, while the scale factor is in parts per million (ppm). The rotation values are defined in two different ways. It's possible to define the rotation angles as positive either clockwise or counterclockwise as you look toward the origin of the X,Y,Z systems.

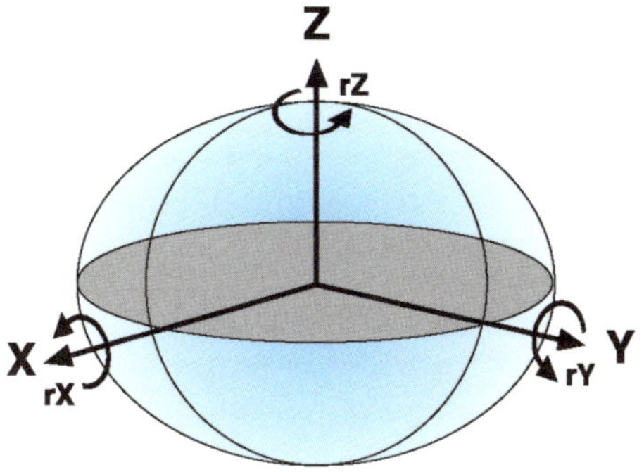

The Coordinate Frame (or Bursa–Wolf) definition of the rotation values.

The equation in the previous column is how the United States and Australia define the equations and is called the Coordinate Frame Rotation transformation. The rotations are positive counterclockwise. Europe uses a different convention called the Position Vector transformation. Both methods are sometimes referred to as the Bursa–Wolf method. In the Projection Engine, the Coordinate Frame and Bursa–Wolf methods are the same. Both Coordinate Frame and Position Vector methods are supported, and it is easy to convert transformation values from one method to the other simply by changing the signs of the three rotation values. For example, the parameters to convert from the WGS 1972 datum to the WGS 1984 datum with the Coordinate Frame method are (in the order, ΔX, ΔY,ΔZ,r_x,r_y,r_z,s):

```
(0.0, 0.0, 4.5, 0.0, 0.0, -0.554,
0.227)
```

To use the same parameters with the Position Vector method, change the sign of the rotation so the new parameters are:

```
(0.0, 0.0, 4.5, 0.0, 0.0, +0.554, 0.227)
```

Unless explicitly stated, it's impossible to tell from the parameters alone which convention is being used. If you use the wrong method, your results can return inaccurate coordinates. The only way to determine how the parameters are defined is by checking a control point whose coordinates are known in the two systems.

Molodensky method

The Molodensky method converts directly between two geographic coordinate systems without actually converting to an X,Y,Z system. The Molodensky method requires three shifts (ΔX,ΔY,ΔZ) and the differences between the semimajor axes (Δa) and the flattenings (Δf) of the two spheroids. The Projection Engine automatically calculates the spheroid differences according to the datums involved.

$$(M + h)\Delta\varphi = -\sin\varphi\cos\lambda\Delta X - \sin\varphi\sin\lambda\Delta Y + \cos\varphi\Delta Z + \frac{e^2\sin\varphi\cos\varphi}{(1-e^2\sin^2\varphi)^{1/2}}\Delta a + \sin\varphi\cos\varphi(M\frac{a}{b} + N\frac{b}{a})\Delta f$$

$$(N + h)\cos\varphi\Delta\lambda = -\sin\lambda\Delta X + \cos\lambda\Delta Y$$

$$\Delta h = \cos\varphi\cos\lambda\,\Delta X + \cos\varphi\sin\lambda\,\Delta Y + \sin\varphi\,\Delta Z - (1-e^2\sin^2\varphi)^{1/2}\Delta a + \frac{a(1-f)}{(1-e^2\sin^2\varphi)^{1/2}}\sin^2\varphi\,\Delta f$$

- h ellipsoid height (meters)
- φ latitude
- λ longitude
- a semimajor axis of the spheroid (meters)
- b semiminor axis of the spheroid (meters)
- f flattening of the spheroid
- e eccentricity of the spheroid

M and N are the meridional and prime vertical radii of curvature, respectively, at a given latitude. The equations for M and N are:

$$M = \frac{a(1-e^2)}{(1-e^2\sin^2\varphi)^{3/2}}$$

$$N = \frac{a}{(1-e^2\sin^2\varphi)^{1/2}}$$

You solve for Δλ and Δφ. The amounts are added automatically by the Projection Engine.

Abridged Molodensky method

The Abridged Molodensky method is a simplified version of the Molodensky method. The equations are:

$$M\Delta\varphi = -\sin\varphi\cos\lambda\Delta X - \sin\varphi\sin\lambda\Delta Y + \cos\varphi\Delta Z + (a\Delta f + f\Delta a)\cdot 2\sin\varphi\cos\varphi$$

$$N\cos\varphi\Delta\lambda = -\sin\lambda\Delta X + \cos\lambda\Delta Y$$

$$\Delta h = \cos\varphi\cos\lambda\Delta X + \cos\varphi\sin\lambda\Delta Y + \sin\varphi\Delta Z + (a\Delta f + f\Delta a)\sin^2\varphi - \Delta a$$

NADCON and HARN methods

The United States uses a grid-based method to convert between geographic coordinate systems. Grid-based methods allow you to model the differences between the systems and are potentially the most accurate method. The area of interest is divided into cells. The National Geodetic Survey (NGS) publishes grids to convert between NAD 1927 and other older geographic coordinate systems and NAD 1983. We group these transformations into the NADCON method. The main NADCON grid, CONUS, converts the contiguous 48 states. The other NADCON grids convert older geographic coordinate systems to NAD 1983 for

- Alaska
- Hawaiian islands
- Puerto Rico and Virgin Islands
- St. George, St. Lawrence, and St. Paul Islands in Alaska

The accuracy is around 0.15 meters for the contiguous states, 0.50 for Alaska and its islands, 0.20 for Hawaii, and 0.05 for Puerto Rico and the Virgin Islands. Accuracies can vary depending on how good the geodetic data in the area was when the grids were computed (NADCON, 1999).

The Hawaiian islands were never on NAD 1927. They were mapped using several datums that are collectively known as the Old Hawaiian datums.

New surveying and satellite measuring techniques have allowed NGS and the states to update the geodetic control point networks. As each state is finished, the NGS publishes a grid that converts between NAD 1983 and the more accurate control point coordinates. Originally, this effort was called the High Precision Geodetic Network (HPGN). It is now called the High Accuracy Reference Network (HARN). More than 40 states have published HARN grids as of September 2000. HARN transformations have an accuracy around 0.05 meters (NADCON, 2000).

The difference values in decimal seconds are stored in two files: one for longitude and the other for latitude. A bilinear interpolation is used to calculate the exact difference between the two geographic coordinate systems at a point. The grids are binary files, but a program, NADGRD, from the NGS allows you to convert the grids to an American Standard Code for Information Interchange (ASCII) format. Shown at the bottom of the page is the header and first row of the CSHPGN.LOA file. This is the longitude grid for Southern California. The format of the first row of numbers is, in order, the number of columns, number of rows, number of z values (always one), minimum longitude, cell size, minimum latitude, cell size, and not used.

The next 37 values (in this case) are the longitude shifts from -122° to -113° at 32° N in 0.25° intervals in longitude.

```
NADCON EXTRACTED REGION                                          NADGRD
  37   21    1   -122.00000         .25000      32.00000         .25000         .00000
       .007383        .004806        .002222       -.000347       -.002868       -.005296
      -.007570       -.009609       -.011305       -.012517       -.013093       -.012901
      -.011867       -.009986       -.007359       -.004301       -.001389        .001164
       .003282        .004814        .005503        .005361        .004420        .002580
       .000053       -.002869       -.006091       -.009842       -.014240       -.019217
      -.025104       -.035027       -.050254       -.072636       -.087238       -.099279
      -.110968
```

A portion of a HARN grid file.

National Transformation version 2

Like the United States, Canada uses a grid-based method to convert between NAD 1927 and NAD 1983. The National Transformation version 2 (NTv2) method is quite similar to NADCON. A set of binary files contains the differences between the two geographic coordinate systems. A bilinear interpolation is used to calculate the exact values for a point.

Unlike NADCON, which can only use one grid at a time, NTv2 is designed to check multiple grids for the most accurate shift information. A set of low-density base grids exists for Canada. Certain areas such as cities have high-density local subgrids that overlay portions of the base, or parent, grids. If a point is within one of the high-density grids, NTv2 will use the high-density grid; otherwise, the point 'falls through' to the low-density grid.

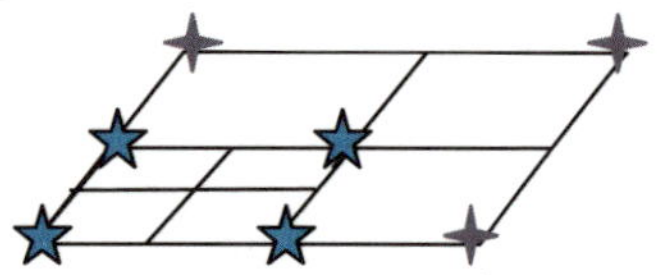

A high-density subgrid with four cells overlaying a low-density base grid, also with four cells.

If a point falls in the lower-left part of the above picture between the stars, the shifts are calculated with the high-density subgrid. A point whose coordinates are anywhere else will have its shifts calculated with the low-density base grid. The software automatically calculates which base or subgrid to use.

The parent grids for Canada have spacings ranging from five to 20 minutes. The high-density grids usually have cell sizes of 30 seconds.

Unlike NADCON grids, NTv2 grids list the accuracy of each point. Accuracy values can range from a few centimeters to around a meter. The high-density grids usually have subcentimeter accuracy.

Australia and New Zealand adopted the NTv2 format to convert between datums as well. Australia has released grids that convert between either Australian Geodetic Datum of 1966 (AGD 1966) or AGD 1984 and Geocentric Datum of Australia of 1994 (GDA 1994). New Zealand has released a countrywide grid to convert between New Zealand Geodetic Datum of 1949 (NZGD 1949) and NZGD 2000.

National Transformation version 1

Like NADCON, the National Transformation version 1 (NTv1) uses a single grid to model the differences between NAD 1927 and NAD 1983 in Canada. This version is also known as CNT in ArcInfo™ Workstation. The accuracy is within 0.01 m of the actual difference for 74 percent of the points and within 0.5 m for 93 percent of the cases.

4 Supported map projections

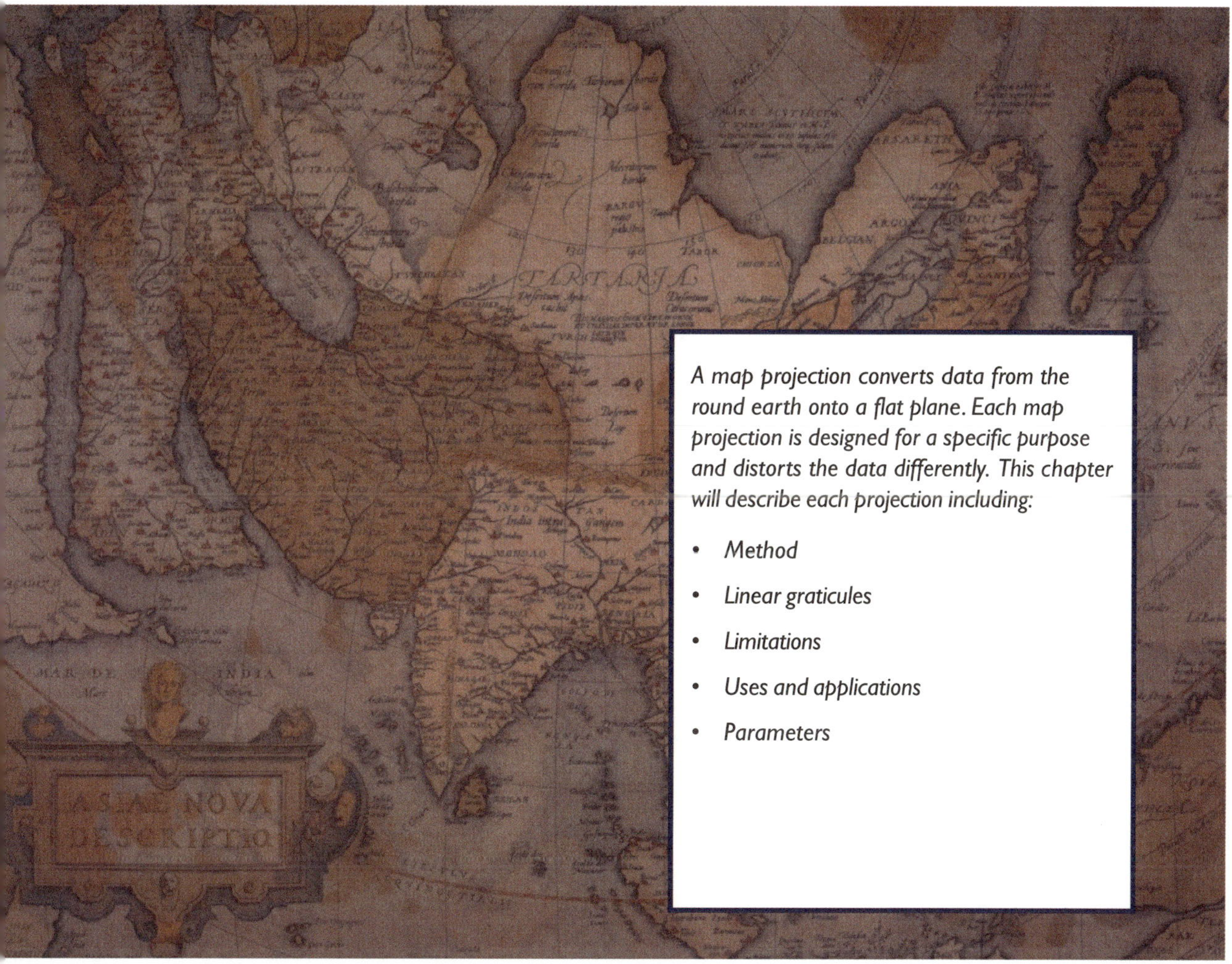

A map projection converts data from the round earth onto a flat plane. Each map projection is designed for a specific purpose and distorts the data differently. This chapter will describe each projection including:

- *Method*
- *Linear graticules*
- *Limitations*
- *Uses and applications*
- *Parameters*

Aitoff	A compromise projection developed in 1889 and used for world maps.
Alaska Grid	This projection was developed to provide a conformal map of Alaska with less scale distortion than other conformal projections.
Alaska Series E	Developed in 1972 by the United States Geological Survey (USGS) to publish a map of Alaska at 1:2,500,000 scale.
Albers Equal Area Conic	This conic projection uses two standard parallels to reduce some of the distortion of a projection with one standard parallel. Shape and linear scale distortion are minimized between the standard parallels.
Azimuthal Equidistant	The most significant characteristic of this projection is that both distance and direction are accurate from the central point.
Behrmann Equal Area Cylindrical	This projection is an equal-area cylindrical projection suitable for world mapping.
Bipolar Oblique Conformal Conic	This projection was developed specifically for mapping North and South America and maintains conformality.
Bonne	This equal-area projection has true scale along the central meridian and all parallels.
Cassini–Soldner	This transverse cylindrical projection maintains scale along the central meridian and all lines parallel to it. This projection is neither equal area nor conformal.
Chamberlin Trimetric	This projection was developed and used by the National Geographic Society for continental mapping. The distance from three input points to any other point is approximately correct.
Craster Parabolic	This pseudocylindrical equal-area projection is primarily used for thematic maps of the world.
Cube	Projects the world to a box that is then unfolded into a plane.
Cylindrical Equal Area	Lambert first described this equal-area projection in 1772. It is used infrequently.
Double Stereographic	This azimuthal projection is conformal.
Eckert I	This pseudocylindrical projection is used primarily as a novelty map.
Eckert II	A pseudocylindrical equal-area projection.
Eckert III	This pseudocylindrical projection is used primarily for world maps.
Eckert IV	This equal-area projection is used primarily for world maps.
Eckert V	This pseudocylindrical projection is used primarily for world maps.
Eckert VI	This equal-area projection is used primarily for world maps.

Equidistant Conic	This conic projection can be based on one or two standard parallels. As the name implies, all circular parallels are spaced evenly along the meridians.
Equidistant Cylindrical	One of the easiest projections to construct because it forms a grid of equal rectangles.
Equirectangular	This projection is very simple to construct because it forms a grid of equal rectangles.
Fuller	The Fuller projection was created by Buckminster Fuller in 1954. Using an icosahedron, the shape is flattened so that the land masses are not interrupted.
Gall's Stereographic	The Gall's Stereographic projection is a cylindrical projection designed around 1855 with two standard parallels at latitudes 45° N and 45° S.
Gauss–Krüger	This projection is similar to the Mercator except that the cylinder is tangent along a meridian instead of the equator. The result is a conformal projection that does not maintain true directions.
Geocentric Coordinate System	The geocentric coordinate system is not a map projection. The earth is modeled as a sphere or spheroid in a right-handed X,Y,Z system.
Geographic Coordinate System	The geographic coordinate system is not a map projection. The earth is modeled as a sphere or spheroid.
Gnomonic	This azimuthal projection uses the center of the earth as its perspective point.
Great Britain National Grid	This coordinate system uses a Transverse Mercator projected on the Airy spheroid. The central meridian is scaled to 0.9996. The origin is 49° N and 2° W.
Hammer–Aitoff	The Hammer–Aitoff projection is a modification of the Lambert Azimuthal Equal Area projection.
Hotine Oblique Mercator	This is an oblique rotation of the Mercator projection. Developed for conformal mapping of areas that do not follow a north–south or east–west orientation but are obliquely oriented.
Krovak	The Krovak projection is an oblique Lambert conformal conic projection designed for the former Czechoslovakia.
Lambert Azimuthal Equal Area	This projection preserves the area of individual polygons while simultaneously maintaining true directions from the center.
Lambert Conformal Conic	This projection is one of the best for middle latitudes. It is similar to the Albers Conic Equal Area projection except that the Lambert Conformal Conic projection portrays shape more accurately than area.
Local Cartesian Projection	This is a specialized map projection that does not take into account the curvature of the earth.

Loximuthal	This projection shows loxodromes, or rhumb lines, as straight lines with the correct azimuth and scale from the intersection of the central meridian and the central parallel.
McBryde–Thomas Flat-Polar Quartic	This equal-area projection is primarily used for world maps.
Mercator	Originally created to display accurate compass bearings for sea travel. An additional feature of this projection is that all local shapes are accurate and clearly defined.
Miller Cylindrical	This projection is similar to the Mercator projection except that the polar regions are not as areally distorted.
Mollweide	Carl B. Mollweide created this pseudocylindrical projection in 1805. It is an equal-area projection designed for small-scale maps.
New Zealand National Grid	This is the standard projection for large-scale maps of New Zealand.
Orthographic	This perspective projection views the globe from an infinite distance. This gives the illusion of a three-dimensional globe.
Perspective	This projection is similar to the Orthographic projection in that its perspective is from space. In this projection, the perspective point is not an infinite distance away; instead, you can specify the distance.
Plate Carrée	This projection is very simple to construct because it forms a grid of equal rectangles.
Polar Stereographic	The projection is equivalent to the polar aspect of the Stereographic projection on a spheroid. The central point is either the North Pole or the South Pole.
Polyconic	The name of this projection translates into 'many cones' and refers to the projection methodology.
Quartic Authalic	This pseudocylindrical equal-area projection is primarily used for thematic maps of the world.
Rectified Skewed Orthomorphic	This oblique cylindrical projection is provided with two options for the national coordinate systems of Malaysia and Brunei.
Robinson	A compromise projection used for world maps.
Simple Conic	This conic projection can be based on one or two standard parallels.
Sinusoidal	As a world map, this projection maintains equal area despite conformal distortion.
Space Oblique Mercator	This projection is nearly conformal and has little scale distortion within the sensing range of an orbiting mapping satellite such as Landsat.
State Plane Coordinate System (SPCS)	The State Plane Coordinate System is not a projection. It is a coordinate system that divides the 50 states of the United States, Puerto Rico, and the U.S. Virgin Islands into more than 120 numbered sections, referred to as zones.

Stereographic	This azimuthal projection is conformal.
Times	The Times projection was developed by Moir in 1965 for Bartholomew Ltd., a British mapmaking company. It is a modified Gall's Stereographic, but the Times has curved meridians.
Transverse Mercator	Similar to the Mercator except that the cylinder is tangent along a meridian instead of the equator. The result is a conformal projection that does not maintain true directions.
Two-Point Equidistant	This modified planar projection shows the true distance from either of two chosen points to any other point on a map.
Universal Polar Stereographic (UPS)	This form of the Polar Stereographic maps areas north of 84° N and south of 80° S that are not included in the UTM Coordinate System. The projection is equivalent to the polar aspect of the Stereographic projection of the spheroid with specific parameters.
Universal Transverse Mercator (UTM)	The Universal Transverse Mercator coordinate system is a specialized application of the Transverse Mercator projection. The globe is divided into 60 zones, each spanning six degrees of longitude.
Van Der Grinten I	This projection is similar to the Mercator projection except that it portrays the world as a circle with a curved graticule.
Vertical Near-Side Perspective	Unlike the Orthographic projection, this perspective projection views the globe from a finite distance. This perspective gives the overall effect of the view from a satellite.
Winkel I	A pseudocylindrical projection used for world maps that averages the coordinates from the Equirectangular (Equidistant Cylindrical) and Sinusoidal projections.
Winkel II	A pseudocylindrical projection that averages the coordinates from the Equirectangular and Mollweide projections.
Winkel Tripel	A compromise projection used for world maps that averages the coordinates from the Equirectangular (Equidistant Cylindrical) and Aitoff projections

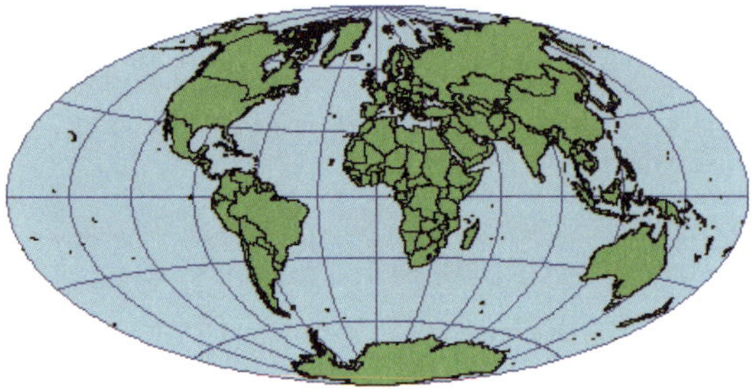

The central meridian is 0°.

DESCRIPTION

A compromise projection developed in 1889 and for use with world maps.

PROJECTION METHOD

Modified azimuthal. Meridians are equally spaced and concave toward the central meridian. The central meridian is a straight line and half the length of the equator. Parallels are equally spaced curves, concave toward the poles.

LINEAR GRATICULES

The equator and the central meridian.

PROPERTIES

Shape

Distortion is moderate.

Area

Moderate distortion.

Direction

Generally distorted.

Distance

The equator and central meridian are at true scale.

LIMITATIONS

Neither conformal nor equal area. Useful only for world maps.

USES AND APPLICATIONS

Developed for use in general world maps.

Used for the Winkel Tripel projection.

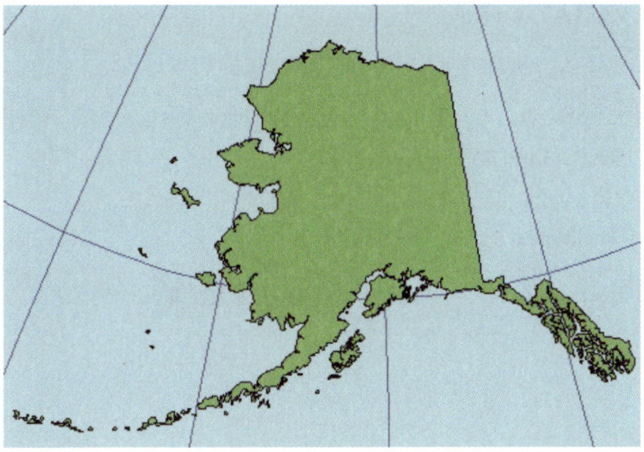

Parameters are set by the software.

DESCRIPTION

This projection was developed to provide a conformal map of Alaska with less scale distortion than other conformal projections. A set of mathematical formulas defines a conformal transformation between two surfaces (Snyder, 1993).

PROJECTION METHOD

Modified planar. This is a sixth-order equation modification of an oblique Stereographic conformal projection on the Clarke 1866 spheroid. The origin is at 64° N, 152° W.

POINT OF TANGENCY

Conceptual point of tangency at 64° N, 152° W.

LINEAR GRATICULES

None.

PROPERTIES

Shape

Perfectly conformal.

Area

Varies about 1.2 percent over Alaska.

Direction

Local angles are correct everywhere.

Distance

The minimum scale factor is 0.997 at approximately 62°30' N, 156° W. Scale increases outward from this point. Most of Alaska and the Aleutian Islands, excluding the panhandle, are bounded by a line of true scale. The scale factor ranges from 0.997 to 1.003 for Alaska, which is one-fourth the range for a corresponding conic projection (Snyder, 1987).

LIMITATIONS

Distortion becomes severe away from Alaska.

USES AND APPLICATIONS

Conformal mapping of Alaska as a complete state on the Clarke 1866 spheroid or NAD27. This projection is not optimized for use with other datums and spheroids.

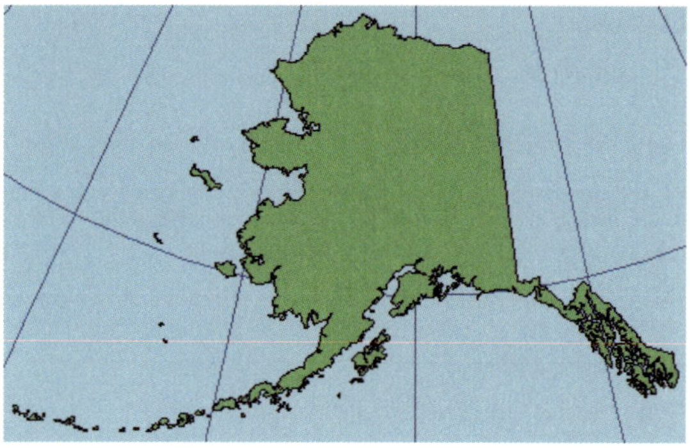

Parameters are set by the software.

DESCRIPTION

This projection was developed in 1972 by the USGS to publish a map of Alaska at 1:2,500,000 scale.

PROJECTION METHOD

Approximates Equidistant Conic, although it is commonly referred to as a Modified Transverse Mercator.

LINES OF CONTACT

The standard parallels at 53°30' N and 66°05'24" N.

LINEAR GRATICULES

The meridians are straight lines radiating from a center point. The parallels closely approximate concentric circular arcs.

PROPERTIES

Shape

Neither conformal nor equal area.

Area

Neither conformal nor equal area.

Direction

Distortion increases with distance from the standard parallels.

Distance

Accurate along the standard parallels.

LIMITATIONS

This projection is appropriate for mapping Alaska, the Aleutian Islands, and the Bering Sea region only.

USES AND APPLICATIONS

1972 USGS revision of a 1954 Alaska map that was published at 1:2,500,000 scale.

1974 map of the Aleutian Islands and the Bering Sea.

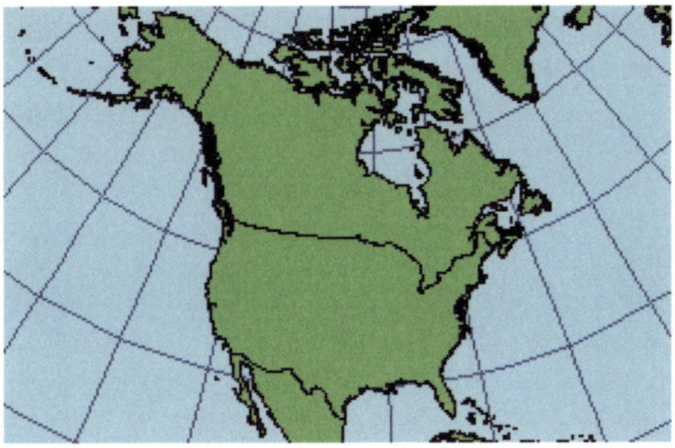

The central meridian is 96° W. The first and second standard parallels are 20° N and 60° N, while the latitude of origin is 40° N.

DESCRIPTION

This conic projection uses two standard parallels to reduce some of the distortion of a projection with one standard parallel. Although neither shape nor linear scale is truly correct, the distortion of these properties is minimized in the region between the standard parallels. This projection is best suited for land masses extending in an east-to-west orientation rather than those lying north to south.

PROJECTION METHOD

Conic. The meridians are equally spaced straight lines converging to a common point. Poles are represented as arcs rather than as single points. Parallels are unequally spaced concentric circles whose spacing decreases toward the poles.

LINES OF CONTACT

Two lines, the standard parallels, defined by degrees latitude.

LINEAR GRATICULES

All meridians.

PROPERTIES

Shape

Shape along the standard parallels is accurate and minimally distorted in the region between the standard parallels and those regions just beyond. The 90 degree angles between meridians and parallels are preserved, but because the scale along the lines of longitude does not match the scale along the lines of latitude, the final projection is not conformal.

Area

All areas are proportional to the same areas on the earth.

Direction

Locally true along the standard parallels.

Distance

Distances are most accurate in the middle latitudes. Along parallels, scale is reduced between the standard parallels and increased beyond them. Along meridians, scale follows an opposite pattern.

LIMITATIONS

Best results for regions predominantly east–west in orientation and located in the middle latitudes. Total range in latitude from north to south should not exceed 30–35 degrees. No limitations on the east–west range.

USES AND APPLICATIONS

Used for small regions or countries but not for continents.

Used for the conterminous United States, normally using 29°30' and 45°30' as the two standard parallels. For this projection, the maximum scale distortion for the 48 states is 1.25 percent.

One method to calculate the standard parallels is by determining the range in latitude in degrees north to south and dividing this range by six. The 'One-Sixth Rule' places the first standard parallel at one sixth the range above the southern boundary and the second standard parallel minus one-sixth the range below the northern limit. There are other possible approaches.

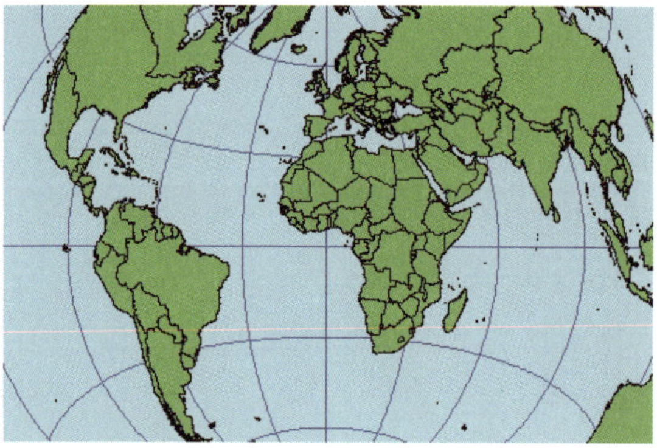

The center of the projection is 0°, 0°.

DESCRIPTION

The most significant characteristic is that both distance and direction are accurate from the central point. This projection can accommodate all aspects: equatorial, polar, and oblique.

PROJECTION METHOD

Planar. The world is projected onto a flat surface from any point on the globe. Although all aspects are possible, the one used most commonly is the polar aspect, in which all meridians and parallels are divided equally to maintain the equidistant property. Oblique aspects centered on a city are also common.

POINT OF TANGENCY

A single point, usually the North or the South Pole, defined by degrees of latitude and longitude.

LINEAR GRATICULES

Polar—Straight meridians are divided equally by concentric circles of latitude.

Equatorial—The equator and the projection's central meridian are linear and meet at a 90-degree angle.

Oblique—The central meridian is straight, but there are no 90-degree intersections except along the central meridian.

PROPERTIES

Shape

Except at the center, all shapes are distorted. Distortion increases from the center.

Area

Distortion increases outward from the center point.

Direction

True directions from the center outward.

Distance

Distances for all aspects are accurate from the center point outward. For the polar aspect, the distances along the meridians are accurate, but there is a pattern of increasing distortion along the circles of latitude, outward from the center.

LIMITATIONS

Usually limited to 90 degrees from the center, although it can project the entire globe. Polar-aspect projections are best for regions within a 30 degree radius because there is only minimal distortion.

Degrees from center:				
15	30	45	60	90
Scale distortion in percent along parallels:				
1.2	4.7	11.1	20.9	57

USES AND APPLICATIONS

Routes of air and sea navigation. These maps will focus on an important location as their central point and use an appropriate aspect.

Polar aspect—Polar regions and polar navigation.

Equatorial aspect—Locations on or near the equator, such as Singapore.

Oblique aspect—Locations between the poles and the equator; for example, large-scale mapping of Micronesia.

If this projection is used on the entire globe, the immediate hemisphere can be recognized and resembles the Lambert Azimuthal projection. The outer hemisphere greatly distorts shapes and areas. In the extreme, a polar-aspect projection centered on the North Pole will represent the South Pole as its largest outermost circle. The function of this extreme projection is that, regardless of the conformal and area distortion, an accurate presentation of distance and direction from the center point is maintained.

BEHRMANN EQUAL AREA CYLINDRICAL

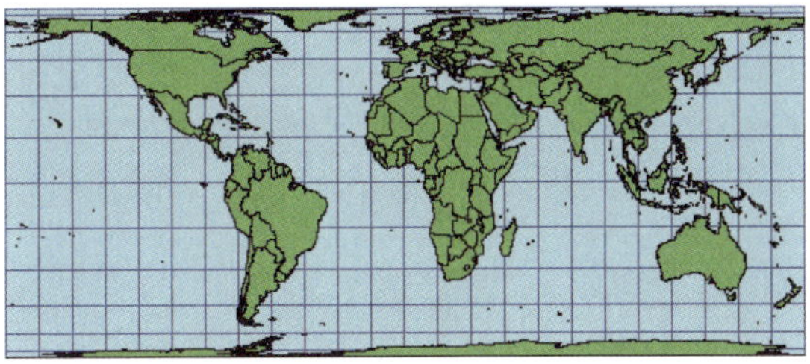

The central meridian is 0°.

DESCRIPTION

This projection is an equal-area cylindrical projection suitable for world mapping.

PROJECTION METHOD

Cylindrical. Standard parallels are at 30° N and S. A case of Cylindrical Equal Area.

LINES OF CONTACT

The two parallels at 30° N and S.

LINEAR GRATICULES

Meridians and parallels are linear.

PROPERTIES

Shape

Shape distortion is minimized near the standard parallels. Shapes are distorted north–south between the standard parallels and distorted east–west above 30° N and below 30° S.

Area

Area is maintained.

Direction

Directions are generally distorted.

Distance

Directions are generally distorted except along the equator.

LIMITATIONS

Useful for world maps only.

USES AND APPLICATIONS

Only useful for world maps.

DESCRIPTION

This projection was developed specifically for mapping North and South America. It maintains conformality. It is based on the Lambert Conformal Conic, using two oblique conic projections side by side.

PROJECTION METHOD

Two oblique conics are joined with the poles 104 degrees apart. A great circle arc 104 degrees long begins at 20° S and 110° W, cuts through Central America, and terminates at 45° N and approximately 19°59'36" W. The scale of the map is then increased by approximately 3.5 percent. The origin of the coordinates is 17°15' N, 73°02' W (Snyder, 1993).

LINES OF CONTACT

The two oblique cones are each conceptually secant. These standard lines do not follow any single parallel or meridian.

LINEAR GRATICULES

Only from each transformed pole to the nearest actual pole.

PROPERTIES

Shape

Conformality is maintained except for a slight discrepancy at the juncture of the two conic projections.

Area

Minimal distortion near the standard lines, increasing with distance.

Direction

Local directions are accurate because of conformality.

Distance

True along standard lines.

LIMITATIONS

Specialized for displaying North and South America only, together. The Bipolar Oblique projection will display North America and South America only. If having problems, check all feature types—particularly annotation and tics—and remove any features that are beyond the range of the projection.

USES AND APPLICATIONS

Developed in 1941 by the American Geographical Society as a low-error single map of North and South America.

Conformal mapping of North and South America as a contiguous unit.

Used by USGS for geologic mapping of North America until it was replaced in 1979 by the Transverse Mercator projection.

The central meridian is 0°.

DESCRIPTION

This equal-area projection has true scale along the central meridian and all parallels. Equatorial aspect is a Sinusoidal. Polar aspect is a Werner.

PROJECTION METHOD

Pseudoconic. Parallels of latitude are equally spaced concentric circular arcs, marked true to scale for meridians.

POINT OF TANGENCY

A single standard parallel with no distortion.

LINEAR GRATICULES

The central meridian.

PROPERTIES

Shape

No distortion along the central meridian and standard parallel; error increases away from these lines.

Area

Equal area.

Direction

Locally true along central meridian and standard parallel.

Distance

Scale is true along the central meridian and each parallel.

LIMITATIONS

Usually limited to maps of continents or smaller regions. Distortion pattern makes other equal-area projections preferable.

USES AND APPLICATIONS

Used during the 19th and early 20th century for atlas maps of Asia, Australia, Europe, and North America. Replaced with the Lambert Azimuthal Equal Area projection for continental mapping by Rand McNally & Co. and Hammond, Inc.

Large-scale topographic mapping of France and Ireland, along with Morocco and some other Mediterranean countries (Snyder, 1993).

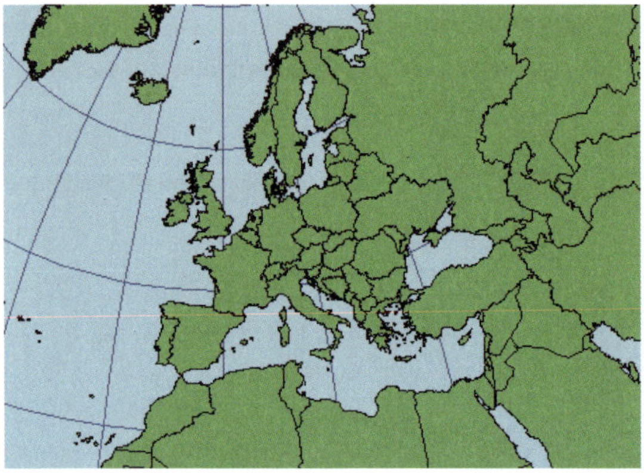

The center of the projection is 0°, 0°.

DESCRIPTION

This transverse cylindrical projection maintains scale along the central meridian and all lines parallel to it and is neither equal area nor conformal. It is most suited for large-scale mapping of areas predominantly north–south in extent. Also called Cassini.

PROJECTION METHOD

A transverse cylinder is projected conceptually onto the globe and is tangent along the central meridian. Cassini–Soldner is analogous to the Equirectangular projection in the same way Transverse Mercator is to the Mercator projection. The name Cassini–Soldner refers to the more accurate ellipsoidal version, developed in the 19th century and used in this software.

POINT OF TANGENCY

Conceptually a line, specified as the central meridian.

LINEAR GRATICULES

The equator, central meridian, and meridians 90 degrees from the central meridian.

PROPERTIES

Shape

No distortion along the central meridian. Distortion increases with distance from the central meridian.

Area

No distortion along the central meridian. Distortion increases with distance from the central meridian.

Direction

Generally distorted.

Distance

Scale distortion increases with distance from the central meridian; however, scale is accurate along the central meridian and all lines perpendicular to the central meridian.

LIMITATIONS

Used primarily for large-scale mapping of areas near the central meridian. The extent on a spheroid is limited to five degrees to either side of the central meridian. Beyond that range, data projected to Cassini–Soldner may not project back to the same position. Transverse Mercator often is preferred due to the difficulty in measuring scale and direction on Cassini–Soldner.

USES AND APPLICATIONS

Normally used for large-scale maps of areas predominantly north–south in extent.

Used for the Ordnance Survey of Great Britain and some German states in the late 19th century. Also used in Cyprus, former Czechoslovakia, Denmark, Malaysia, and the former Federal Republic of Germany.

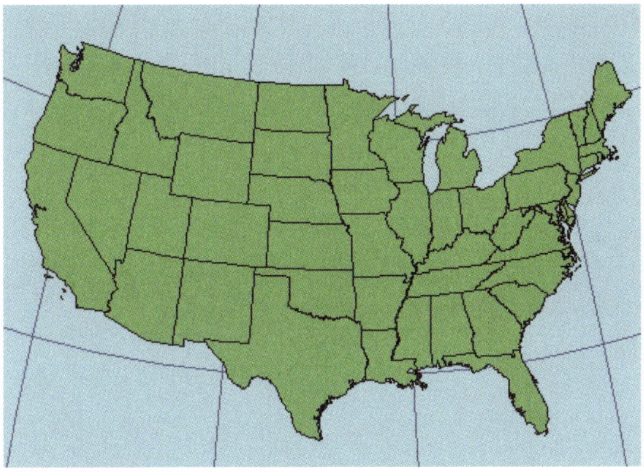

The three points that define the projection are 120° W, 48° N; 98° W, 27° N; and 70° W, 45° N.

DESCRIPTION

This is the standard projection developed and used by the National Geographic Society for continental mapping. The distance from three input points to any other point is approximately correct.

PROJECTION METHOD

Modified planar.

LINEAR GRATICULES

None.

PROPERTIES

Shape

Shape distortion is low throughout if the three points are placed near the map limits.

Area

Areal distortion is low throughout if the three points are placed near the map limits.

Direction

Low distortion throughout.

Distance

Nearly correct representation of distance from three widely spaced points to any other point.

LIMITATIONS

The three selected input points should be widely spaced near the edge of the map limits.

Chamberlin Trimetric can only be used in ArcInfo as an OUTPUT projection because the inverse equations—Chamberlin Trimetric to geographic—have not been published.

You can't project an ArcInfo grid or lattice to Chamberlin Trimetric because the inverse equations are required.

USES AND APPLICATIONS

Used by the National Geographic Society as the standard map projection for most continents.

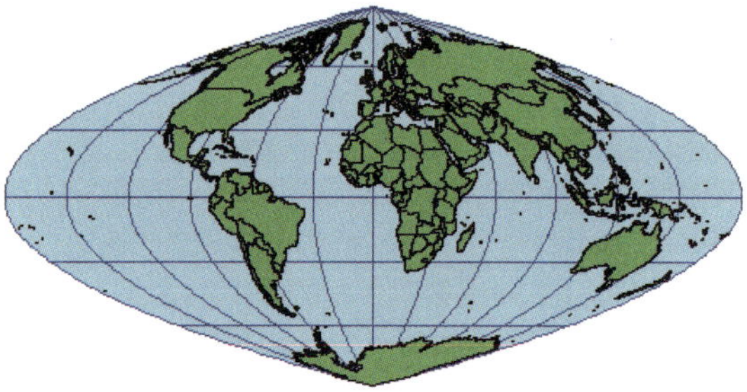

The central meridian is 0°.

DESCRIPTION

This pseudocylindrical equal area projection is primarily used for thematic maps of the world. Also known as Putnins P4.

PROJECTION METHOD

Pseudocylindrical.

LINEAR GRATICULES

The central meridian is a straight line half as long as the equator. Parallels are unequally spaced, straight parallel lines perpendicular to the central meridian. Their spacing decreases very gradually as they move away from the equator.

PROPERTIES

Shape

Free of distortion at the central meridian at 36°46' N and S. Distortion increases with distance from these points and is most severe at the outer meridians and high latitudes. Interrupting the projection greatly reduces this distortion.

Area

Equal area.

Direction

Local angles are correct at the intersection of 36°46' N and S with the central meridian. Direction is distorted elsewhere.

Distance

Scale is true along latitudes 36°46' N and S. Scale is also constant along any given latitude and is symmetrical around the equator.

LIMITATIONS

Useful only as a world map.

USES AND APPLICATIONS

Thematic world maps.

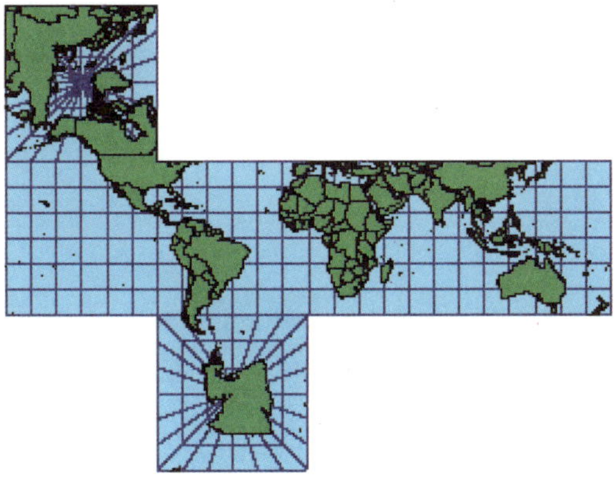

The Cube projection with option 1.

DESCRIPTION

A faceted projection that is used for ArcGlobe™.

PROJECTION METHOD

The world is projected onto a six-sided figure with square sides.

POINTS OF INTERSECTION

The poles.

LINEAR GRATICULES

The meridians are straight lines. Parallels are squares on the two polar facets and straight lines on the equatorial facets.

PROPERTIES

Shape

Shapes are not maintained.

Area

Areas are not maintained.

Direction

Between +45 and -45 degrees latitude, north, south, east, and west directions are accurate, but general directions are not. On the polar facets, directions are true from the poles.

Distance

Between +45 and -45 degrees latitude, scale is correct along the meridians. On the polar facets, only the central longitude line has the correct scale.

LIMITATIONS

Not recommended for general mapping.

USES AND APPLICATIONS

Used for ArcGlobe. Suitable for educational uses.

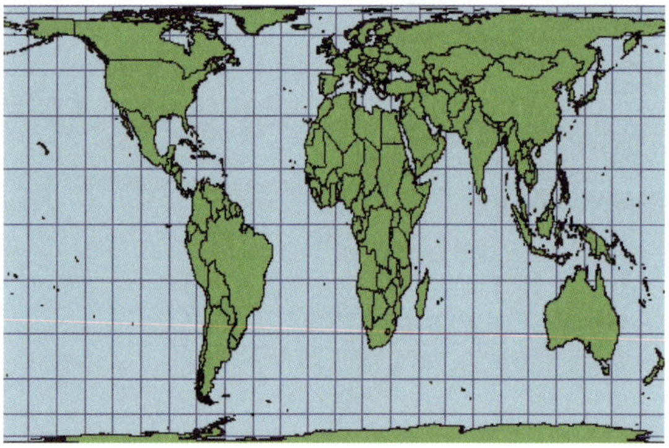

The central meridian is 0°, and the standard parallel is 40° N. The opposite parallel, 40° S, is also a standard parallel.

DESCRIPTION

Lambert first described this equal area projection in 1772. It has been used infrequently.

PROJECTION METHOD

A normal perspective projection onto a cylinder tangent at the equator.

POINTS OF INTERSECTION

The equator.

LINEAR GRATICULES

In the normal, or equatorial aspect, all meridians and parallels are perpendicular straight lines. Meridians are equally spaced and 0.32 times the length of the equator. Parallels are unequally spaced and farthest apart near the equator. Poles are lines of length equal to the equator.

PROPERTIES

Shape

Shape is true along the standard parallels of the normal aspect. Distortion is severe near the poles of the normal aspect.

Area

There is no area distortion.

Direction

Local angles are correct along standard parallels or standard lines. Direction is distorted elsewhere.

Distance

Scale is true along the equator. Scale distortion is severe near the poles.

LIMITATIONS

Recommended for narrow areas extending along the central line. Severe distortion of shape and scale near the poles.

USES AND APPLICATIONS

Suitable for equatorial regions.

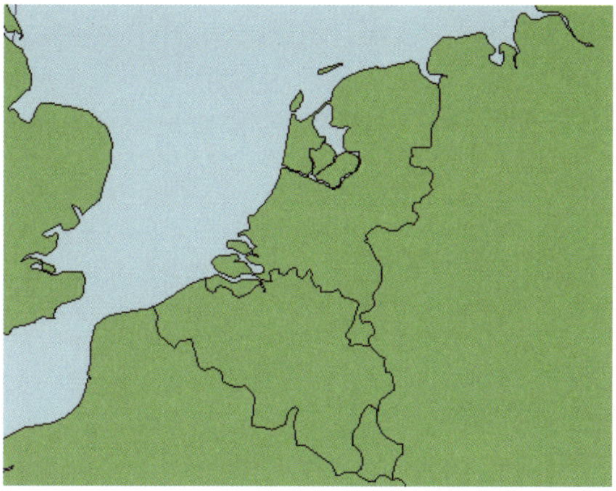

The Rijksdriehoekstelsel coordinate system is used in the Netherlands. The central meridian is 5°23'15.5" E. The latitude of origin is 52°09'22.178" N. The scale factor is 0.9999079. The false easting is 155,000 meters, and the false northing is 463,000 meters.

DESCRIPTION

A conformal projection.

PROJECTION METHOD

Planar perspective projection, viewed from the point on the globe opposite the point of tangency. Points are transformed from the spheroid to a Gaussian sphere before being projected to the plane.

All meridians and parallels are shown as circular arcs or straight lines. Graticular intersections are 90 degrees. In the equatorial aspect, the parallels curve in opposite directions on either side of the equator. In the oblique case, only the parallel with the opposite sign to the central latitude is a straight line; other parallels are concave toward the poles on either side of the straight parallel.

POINT OF CONTACT

A single point anywhere on the globe.

LINEAR GRATICULES

Polar aspect—All meridians.

Equatorial aspect—The central meridian and the equator.

Oblique aspect—The central meridian and parallel of latitude with the opposite sign of the central latitude.

PROPERTIES

Shape

Conformal. Local shapes are accurate.

Area

True scale at center with distortion increasing as you move away from the center.

Direction

Directions are accurate from the center. Local angles are accurate everywhere.

Distance

Scale increases with distance from the center.

LIMITATIONS

Normally limited to one hemisphere. Portions of the outer hemisphere may be shown, but with rapidly increasing distortion.

USES AND APPLICATIONS

Used for large-scale coordinate systems in New Brunswick and the Netherlands.

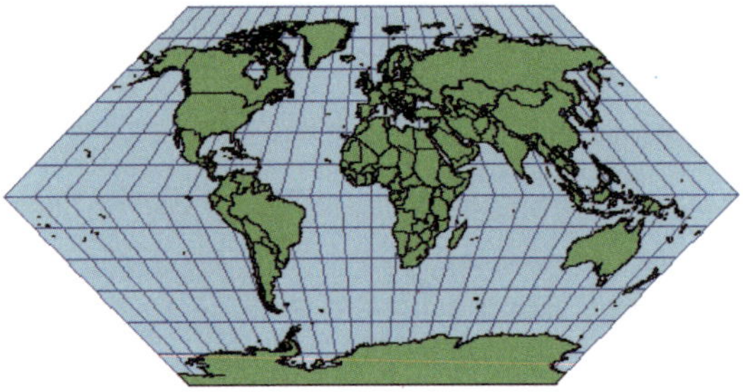

The central meridian is 0°.

DESCRIPTION

Used primarily as a novelty map.

PROJECTION METHOD

A pseudocylindrical projection.

LINEAR GRATICULES

Parallels and meridians are equally spaced straight lines. The poles and the central meridian are straight lines half as long as the equator.

PROPERTIES

Shape

Shape isn't preserved.

Area

Area isn't preserved.

Direction

Direction is distorted everywhere.

Distance

Scale is correct along 47°10' N and S.

LIMITATIONS

Discontinuities exist at the equator.

USES AND APPLICATIONS

Useful only as a novelty.

ECKERT II

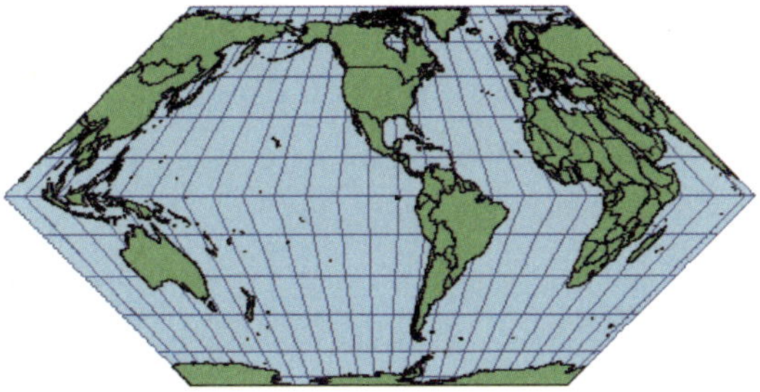

The central meridian is 100° W.

DESCRIPTION

A pseudocylindrical equal-area projection.

PROJECTION METHOD

A pseudocylindrical projection.

Parallels are unequally spaced straight lines. Meridians are equally spaced straight lines. The poles and the central meridian are straight lines half as long as the equator.

PROPERTIES

Shape

Shape isn't preserved.

Area

Area is preserved.

Direction

Direction is distorted everywhere.

Distance

Scale is correct along 55°10' N and S.

LIMITATIONS

Discontinuities exist at the equator.

USES AND APPLICATIONS

Useful only as a novelty.

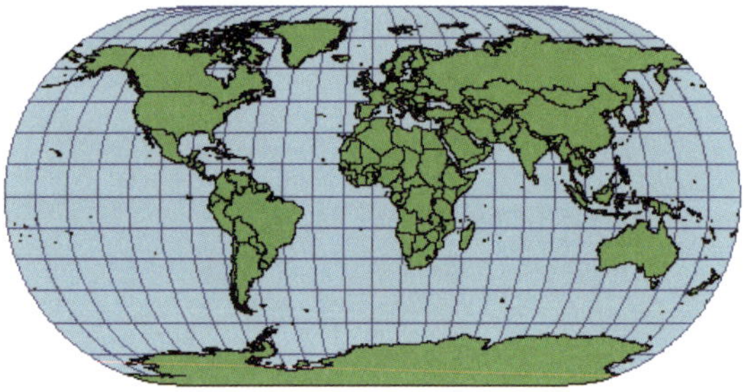

The central meridian is 0°.

DESCRIPTION

This pseudocylindrical projection is used primarily for world maps.

PROJECTION METHOD

A pseudocylindrical projection.

LINEAR GRATICULES

Parallels are equally spaced straight lines. Meridians are equally spaced elliptical curves. The meridians at +/-180° from the central meridian are semicircles. The poles and the central meridian are straight lines half as long as the equator.

PROPERTIES

Shape

This stretching decreases to zero at 37°55' N and S. Nearer the poles, features are compressed in the north–south direction.

Area

Area isn't preserved.

Direction

The equator doesn't have any angular distortion. Direction is distorted elsewhere.

Distance

Scale is correct only along 37°55' N and S. Nearer the poles, features are compressed in the north–south direction.

LIMITATIONS

Useful only as a world map.

USES AND APPLICATIONS

Suitable for thematic mapping of the world.

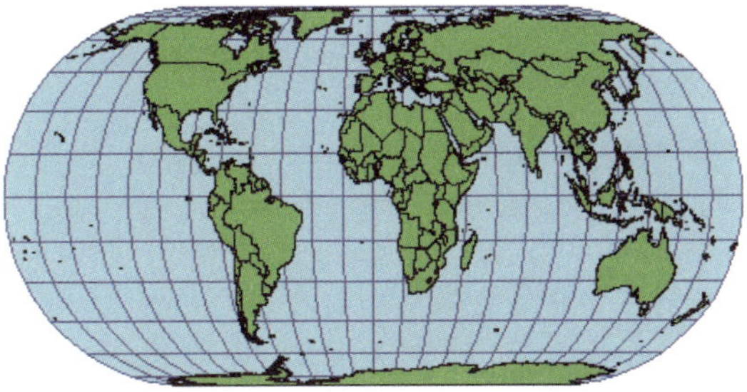

The central meridian is 0°.

DESCRIPTION

This equal area projection is used primarily for world maps.

PROJECTION METHOD

A pseudocylindrical equal-area projection.

LINEAR GRATICULES

Parallels are unequally spaced straight lines, closer together at the poles. Meridians are equally spaced elliptical arcs. The poles and the central meridian are straight lines half as long as the equator.

PROPERTIES

Shape

Shapes are stretched north–south 40 percent along the equator, relative to the east–west dimension. This stretching decreases to zero at 40°30' N and S at the central meridian. Nearer the poles, features are compressed in the north–south direction.

Area

Equivalent.

Direction

Local angles are correct at the intersections of 40°30' N and S with the central meridian. Direction is distorted elsewhere.

Distance

Scale is distorted north–south 40 percent along the equator relative to the east–west dimension. This distortion decreases to zero at 40°30' N and S at the central meridian. Scale is correct only along these parallels. Nearer the poles, features are compressed in the north–south direction.

LIMITATIONS

Useful only as a world map.

USES AND APPLICATIONS

Thematic maps of the world such as climate.

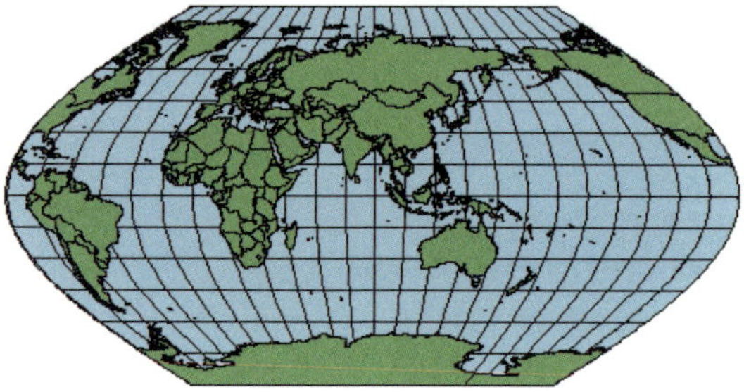

The central meridian is 89° E.

DESCRIPTION

This pseudocylindrical projection is used primarily for world maps.

PROJECTION METHOD

A pseudocylindrical projection.

LINEAR GRATICULES

Parallels are equally spaced straight lines. Meridians are equally spaced sinusoidal curves. The poles and the central meridian are straight lines half as long as the equator.

PROPERTIES

Shape

This stretching decreases to zero at 37°55' N and S. Nearer the poles, features are compressed in the north–south direction.

Area

Area isn't preserved.

Direction

The equator doesn't have any angular distortion. Direction is distorted elsewhere.

Distance

Scale is correct only along 37°55' N and S. Nearer the poles, features are compressed in the north–south direction.

LIMITATIONS

Useful only as a world map.

USES AND APPLICATIONS

Suitable for thematic mapping of the world.

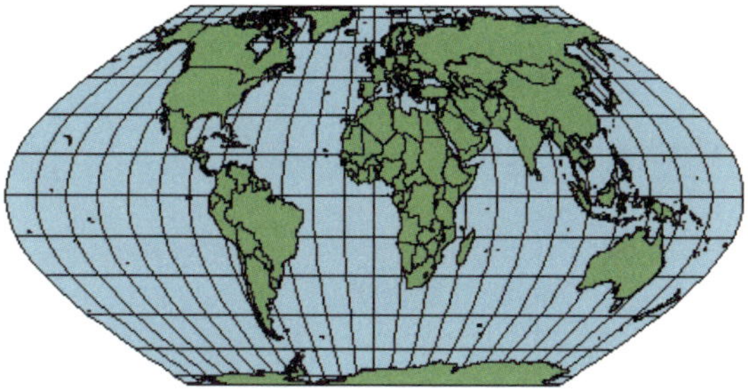

The central meridian is 0°.

DESCRIPTION

This equal-area projection is used primarily for world maps.

PROJECTION METHOD

A pseudocylindrical equal-area projection.

LINEAR GRATICULES

Parallels are unequally spaced straight lines. They are closer together at the poles. Meridians are equally spaced sinusoidal curves. The poles and the central meridian are straight lines half as long as the equator.

PROPERTIES

Shape

Shapes are stretched north–south 29 percent along the equator, relative to the east–west dimension. This stretching decreases to zero at 49°16' N and S at the central meridian. Nearer the poles, features are compressed in the north–south direction.

Area

Equivalent.

Direction

Local angles are correct at the intersection of 49°16' N and S with the central meridian. Direction is distorted elsewhere.

Distance

Scale is distorted north–south 29 percent along the equator relative to the east–west dimension. This distortion decreases to zero at 49°16' N and S at the central meridian. Scale is correct only along these parallels. Nearer the poles, features are compressed in the north–south direction.

LIMITATIONS

Useful only as a world map.

USES AND APPLICATIONS

Suitable for thematic mapping of the world.

Used for world distribution maps in the *1937 World Atlas* by the Soviet Union.

The central meridian is 60° W. The first and second standard parallels are 5° S and 42° S. The latitude of origin is 32° S.

DESCRIPTION

This conic projection can be based on one or two standard parallels. As its name implies, all circular parallels are spaced evenly along the meridians. This is true whether one or two parallels are used as the standards.

PROJECTION METHOD

Cone is tangential if one standard parallel is specified and secant if two standard parallels are specified. Graticules are evenly spaced. Meridian spacing is equal, as is the space between each of the concentric arcs that describe the lines of latitude. The poles are represented as arcs rather than points.

If the pole is given as the single standard parallel, the cone becomes a plane and the resulting projection is the same as a polar Azimuthal Equidistant.

If two standard parallels are placed symmetrically north and south of the equator, the resulting projection is the same as the Equirectangular projection. In this case, you must use the Equirectangular projection.

Use the Equirectangular projection if the standard parallel is the equator.

LINES OF CONTACT

Depends on the number of standard parallels.

Tangential projections (Type 1)—One line, indicated by the standard parallel.

Secant projections (Type 2)—Two lines, specified as the first and second standard parallels.

LINEAR GRATICULES

All meridians.

PROPERTIES

Shape

Local shapes are true along the standard parallels. Distortion is constant along any given parallel but increases with distance from the standard parallels.

Area

Distortion is constant along any given parallel but increases with distance from the standard parallels.

Direction

Locally true along the standard parallels.

Distance

True along the meridians and the standard parallels. Scale is constant along any given parallel but changes from parallel to parallel.

LIMITATIONS

Range in latitude should be limited to 30 degrees.

USES AND APPLICATIONS

Regional mapping of midlatitude areas with a predominantly east–west extent.

Common for atlas maps of small countries.

Used by the former Soviet Union for mapping the entire country.

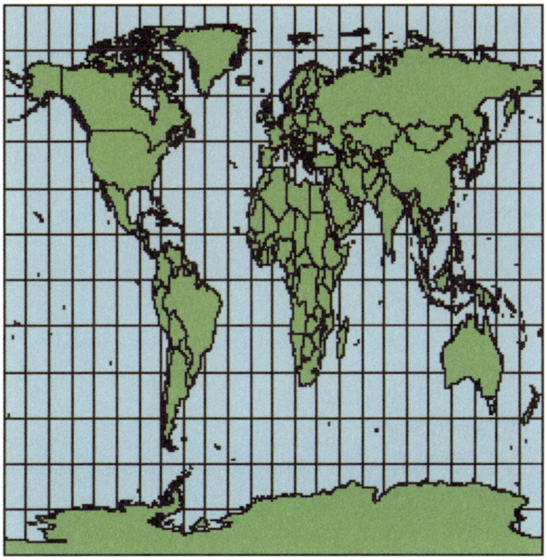

The central meridian is 0°.

DESCRIPTION

Also known as Equirectangular, Simple Cylindrical, Rectangular, or Plate Carrée (if the standard parallel is the equator).

This projection is very simple to construct because it forms a grid of equal rectangles. Because of its simple calculations, its usage was more common in the past. In this projection, the polar regions are less distorted in scale and area than they are in the Mercator projection.

PROJECTION METHOD

This simple cylindrical projection converts the globe into a Cartesian grid. Each rectangular grid cell has the same size, shape, and area. All the graticular intersections are 90 degrees. The central parallel may be any line, but the traditional Plate Carrée projection uses the equator. When the equator is used, the grid cells are perfect squares, but if any other parallel is used, the grids become rectangular. In this projection, the poles are represented as straight lines across the top and bottom of the grid.

LINES OF CONTACT

Tangent at the equator or secant at two parallels symmetrical about the equator.

LINEAR GRATICULES

All meridians and all parallels.

PROPERTIES

Shape

Distortion increases as the distance from the standard parallels increases.

Area

Distortion increases as the distance from the standard parallels increases.

Direction

North, south, east, and west directions are accurate. General directions are distorted, except locally along the standard parallels.

Distance

The scale is correct along the meridians and the standard parallels.

LIMITATIONS

Noticeable distortion of all properties away from standard parallels.

USES AND APPLICATIONS

Best used for city maps or other small areas with map scales large enough to reduce the obvious distortion.

Used for simple portrayals of the world or regions with minimal geographic data. This makes the projection useful for index maps.

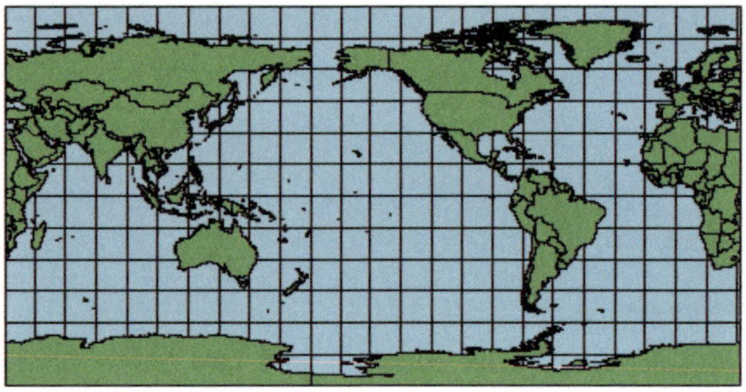

The central meridian is 149° W.

DESCRIPTION

Also known as Simple Cylindrical, Equidistant Cylindrical, Rectangular, or Plate Carrée (if the standard parallel is the equator).

This projection is very simple to construct because it forms a grid of equal rectangles. Because of its simple calculations, its usage was more common in the past. In this projection, the polar regions are less distorted in scale and area than they are in the Mercator projection.

PROJECTION METHOD

This simple cylindrical projection converts the globe into a Cartesian grid. Each rectangular grid cell has the same size, shape, and area. All the graticular intersections are 90 degrees. The central parallel may be any line, but the traditional Plate Carrée projection uses the equator. When the equator is used, the grid cells are perfect squares, but if any other parallel is used, the grids become rectangular. In this projection, the poles are represented as straight lines across the top and bottom of the grid.

LINES OF CONTACT

Tangent at the equator or secant at two parallels symmetrical around the equator.

LINEAR GRATICULES

All meridians and all parallels.

PROPERTIES

Shape

Distortion increases as the distance from the standard parallels increases.

Area

Distortion increases as the distance from the standard parallels increases.

Direction

North, south, east, and west directions are accurate. General directions are distorted, except locally along the standard parallels.

Distance

The scale is correct along the meridians and the standard parallels.

LIMITATIONS

Noticeable distortion of all properties away from standard parallels.

USES AND APPLICATIONS

Best used for city maps or other small areas with map scales large enough to reduce the obvious distortion.

Used for simple portrayals of the world or regions with minimal geographic data. This makes the projection useful for index maps.

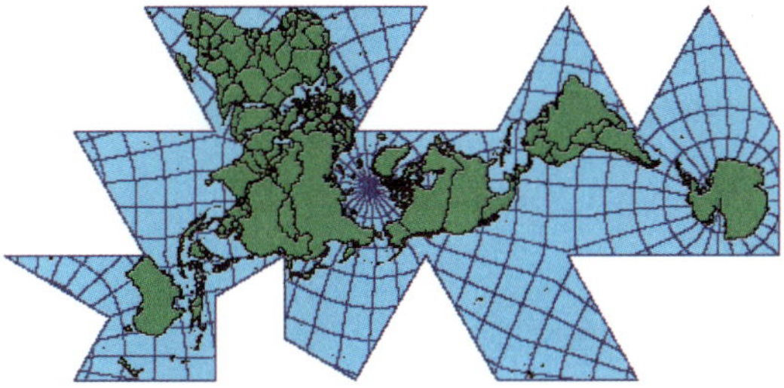

The Fuller projection maintains the cootinuity of the land masses.

DESCRIPTION

The final version was described by Buckminster Fuller in 1954. Contact the Buckminster Fuller Institute at www.bfi.org for more information.

PROJECTION METHOD

This projection converts the globe into a 20-sided figure called an icosahedron. Each side is a geodesic triangle that is then flattened into a two-dimensional triangle. The facets of the icosahedron are unfolded in a specific manner to keep the land masses unbroken.

LINES OF CONTACT

The tangent lines are the facet edges.

LINEAR GRATICULES

In general, neither latitude nor longitude lines are straight.

PROPERTIES

Shape

Distortion increases as the distance from the facet edges increases. Because the Fuller projection is comprised of twenty facets that are projected individually, overall shape distortion is low. Scale is less than one within a facet.

Area

Distortion increases as the distance from the facet edges increases. Scale is less than one within a facet.

Direction

General directions are distorted, depending on the orientation of a facet. Angles within a facet are slightly distorted due to the flattening of the geodesic triangle.

Distance

The scale is correct along the facet edges.

LIMITATIONS

The north direction is not upright. It's difficult to identify directions without a graticule on the map.

USES AND APPLICATIONS

Best used for display and educational uses.

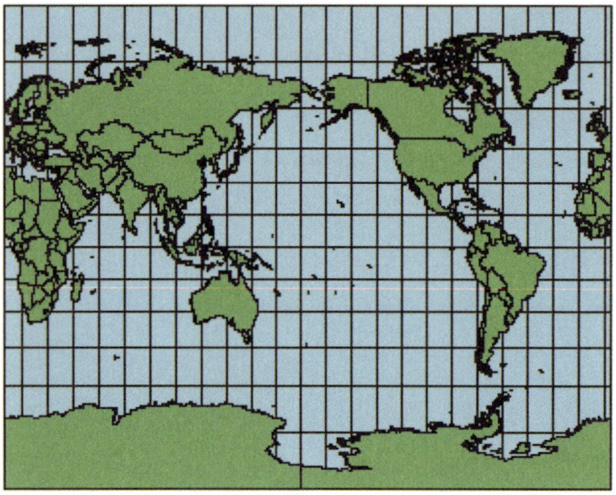

The central meridian is 176° E.

DESCRIPTION

Gall's Stereographic was designed around 1855. It is a cylindrical projection with two standard parallels at latitudes 45° N and S.

PROJECTION METHOD

Cylindrical stereographic projection based on two standard parallels at 45° N and S. The globe is projected perspectively onto a secant cylinder from the point on the equator opposite a given meridian. Meridians are equally spaced straight lines. Parallels are straight lines with spacing increasing away from the equator. Poles are straight lines.

LINES OF CONTACT

Two lines at 45° N and S.

LINEAR GRATICULES

All meridians and parallels.

PROPERTIES

Shape

Shapes are true at latitudes 45° N and S. Distortion slowly increases away from these latitudes and becomes severe at the poles.

Area

Area is true at latitudes 45° N and S. Distortion slowly increases away from these latitudes and becomes severe at the poles.

Direction

Locally correct at latitudes 45° N and S. Generally distorted elsewhere.

Distance

Scale is true in all directions along latitudes 45° N and S. Scale is constant along parallels and is symmetrical around the equator. Distances are compressed between latitudes 45° N and S and expanded beyond them.

LIMITATIONS

Used only for world maps.

USES AND APPLICATIONS

Used for world maps in British atlases.

DESCRIPTION

Also known as Transverse Mercator.

This projection is similar to the Mercator except that the cylinder is longitudinal along a meridian instead of the equator. The result is a conformal projection that does not maintain true directions. The central meridian is placed on the region to be highlighted. This centering minimizes distortion of all properties in that region. This projection is best suited for land masses that stretch north–south. The Gauss–Krüger (GK) coordinate system is based on the Gauss–Krüger projection.

PROJECTION METHOD

Cylindrical projection with central meridian placed in a particular region.

LINES OF CONTACT

Any single meridian for the tangent projection. For the secant projection, two parallel lines equidistant from the central meridian.

LINEAR GRATICULES

The equator and the central meridian.

PROPERTIES

Shape

Conformal. Small shapes are maintained. Shapes of larger regions are increasingly distorted away from the central meridian.

Area

Distortion increases with distance from the central meridian.

Direction

Local angles are accurate everywhere.

Distance

Accurate scale along the central meridian if the scale factor is 1.0. If it is less than 1.0, then there are two straight lines having an accurate scale, equidistant from and on each side of the central meridian.

LIMITATIONS

Data on a spheroid or an ellipsoid cannot be projected beyond 90 degrees from the central meridian. In fact, the extent on a spheroid or ellipsoid should be limited to 10 to 12 degrees on both sides of the central meridian. Beyond that range, data projected may not project back to the same position. Data on a sphere does not have these limitations.

USES AND APPLICATIONS

Gauss–Krüger coordinate system. Gauss–Krüger divides the world into zones six degrees wide. Each zone has a scale factor of 1.0 and a false easting of 500,000 meters. The central meridian of zone 1 is at 3° E. Some places also add the zone number times one million to the 500,000 false easting value. GK zone 5 could have a false easting value of 500,000 or 5,500,000 meters.

The UTM system is very similar. The scale factor is 0.9996, and the central meridian of UTM zone 1 is at 177° W. The false easting value is 500,000 meters, and southern hemisphere zones also have a false northing of 10,000,000.

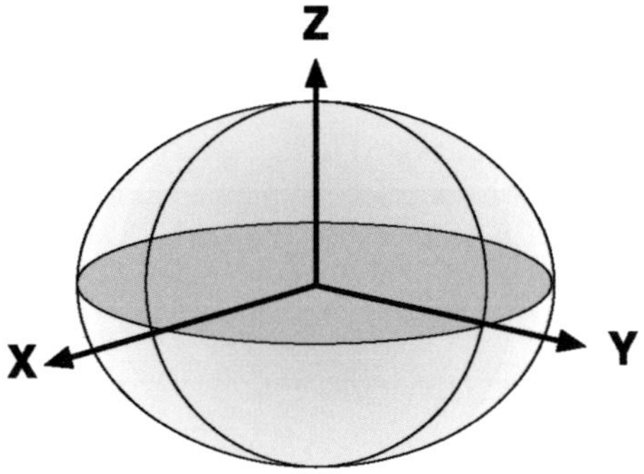

Geographic coordinates are described as X, Y, and Z values in a geocentric coordinate system.

DESCRIPTION

The geocentric coordinate system is not a map projection. The earth is modeled as a sphere or spheroid in a right-handed X,Y,Z system.

The X-axis points to the prime meridian, the Y-axis points 90 degrees away in the equatorial plane, and the Z-axis points in the direction of the North Pole.

USES AND APPLICATIONS

The geocentric coordinate system is used internally as an interim system for several geographic (datum) transformation methods.

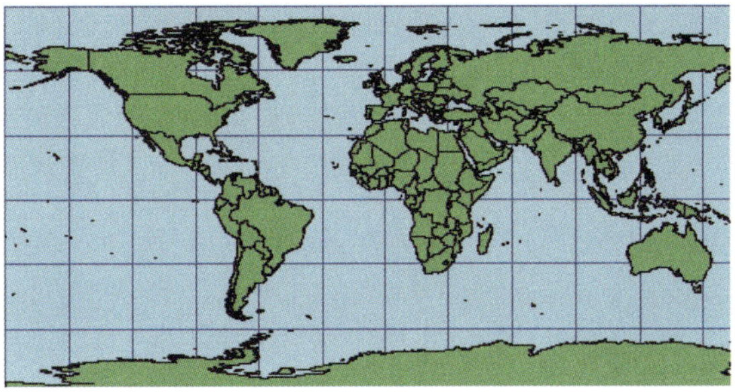

Geographic coordinates displayed as if the longitude–latitude values are linear units. An equivalent projection is Equirectangular with the standard parallel set to the equator.

DESCRIPTION

The geographic coordinate system is not a map projection. The earth is modeled as a sphere or spheroid. The sphere is divided into equal parts, usually called degrees; some countries use grads. A circle is 360 degrees or 400 grads. Each degree is subdivided into 60 minutes, with each minute composed of 60 seconds.

The geographic coordinate system consists of latitude and longitude lines. Each line of longitude runs north–south and measures the number of degrees east or west of the prime meridian. Values range from -180 to +180 degrees. Lines of latitude run east–west and measure the number of degrees north or south of the equator. Values range from +90 degrees at the North Pole to -90 degrees at the South Pole.

The standard origin is where the Greenwich prime meridian meets the equator. All points north of the equator or east of the prime meridian are positive.

USES AND APPLICATIONS

Map projections use latitude and longitude values to reference parameters, such as the central meridian, the standard parallels, and the latitude of origin.

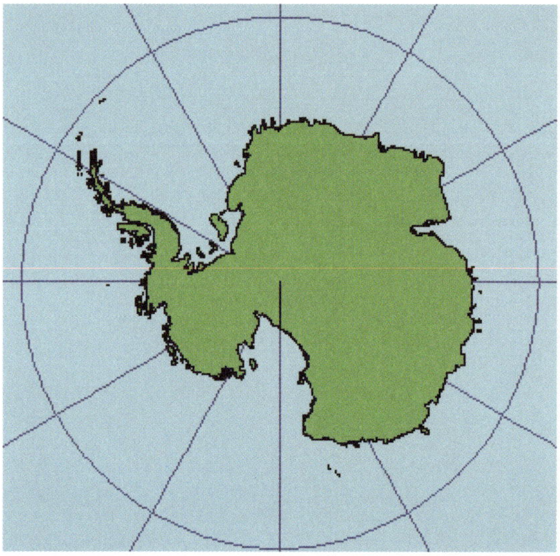

The central meridian is 0°, and the latitude of origin is 90° S.

DESCRIPTION

This azimuthal projection uses the center of the earth as its perspective point. All great circles are straight lines, regardless of the aspect. This is a useful projection for navigation because great circles highlight routes with the shortest distance.

PROJECTION METHOD

This is a planar perspective projection viewed from the center of the globe. The projection can be any aspect.

POINT OF TANGENCY

A single point anywhere on the globe.

Polar aspect—North Pole or South Pole.

Equatorial aspect—Any point along the equator.

Oblique aspect—Any other point.

LINEAR GRATICULES

All meridians and the equator.

PROPERTIES

Shape

Increasingly distorted from the center; moderate distortion within 30 degrees of the center point.

Area

Distortion increases with distance from the center; moderate distortion within a 30-degree radius of the center.

Direction

Accurate from the center.

Distance

No line has an accurate scale, and the amount of distortion increases with distance from the center.

Scalar Distortion for Polar Aspect

Degrees from Center (°)	15.0	30.0	45.0	60.0
Meridian Distortion (%)	7.2	33.3	100.0	300.0
Latitude Distortion (%)	3.5	15.5	41.4	100.0

LIMITATIONS

This projection is limited by its perspective point and cannot project a line that is 90 degrees or more from the center point; this means that the equatorial aspect cannot project the poles and the polar aspects cannot project the equator.

A radius of 30 degrees produces moderate distortion, as indicated in the table above. This projection should not be used more than about 60 degrees from the center.

USES AND APPLICATIONS

All aspects—Routes of navigation for sea and air.

Polar aspect—Navigational maps of polar regions.

Equatorial aspect—Navigational maps of Africa and the tropical region of South America.

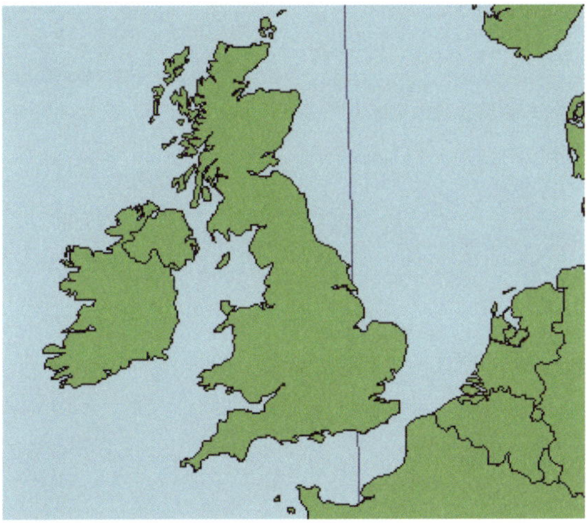

The central meridian is 2° W, and the latitude of origin is 49° N. The scale factor is 0.9996.

DESCRIPTION

This is a Transverse Mercator projected on the Airy spheroid. The central meridian is scaled to 0.9996. The origin is 49° N and 2° W.

PROJECTION METHOD

Cylindrical, transverse projection with the central meridian centered along a particular region.

LINES OF CONTACT

Two lines parallel with and 180 km from the central meridian at 2° W.

LINEAR GRATICULES

The central meridian.

PROPERTIES

Shape

Conformal; therefore, small shapes are maintained accurately.

Area

Distortion increases beyond Great Britain as the distance from the central meridian increases.

Direction

Local directions are accurately maintained.

Distance

Scale is accurate along the lines of secancy 180 km from the central meridian. Scale is compressed between them and expanded beyond them.

LIMITATIONS

Suitable for Great Britain. Limited in east–west extent.

USES AND APPLICATIONS

The national coordinate system for Great Britain; used for large-scale topographic mapping.

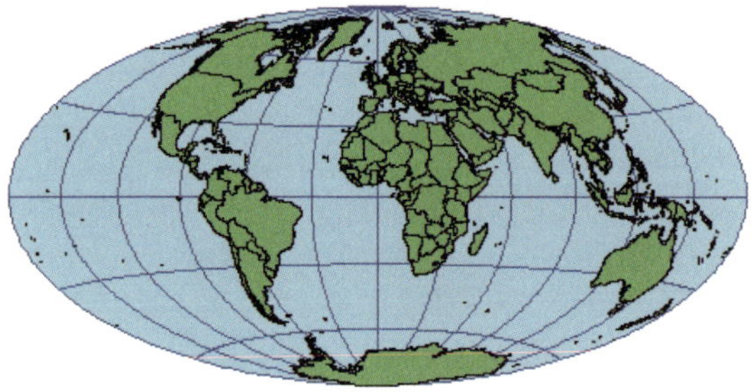

The central meridian is 0°.

DESCRIPTION

The Hammer–Aitoff projection is a modification of the Lambert Azimuthal Equal Area projection.

PROJECTION METHOD

Modified azimuthal. The central meridian is a straight line half as long as the equator. The other meridians are complex curves, concave toward the central meridian and unequally spaced along the equator. The equator is a straight line; all other parallels are complex curves, concave toward the nearest pole and unequally spaced along the central meridian.

POINT OF TANGENCY

Central meridian at the equator.

LINEAR GRATICULES

The equator and central meridian are the only straight lines.

PROPERTIES

Shape

Distortion increases away from the origin.

Area

Equal area.

Direction

Local angles are true only at the center.

Distance

Scale decreases along the equator and central meridian as distance from the origin increases.

LIMITATIONS

Useful only as a world map.

USES AND APPLICATIONS

Thematic maps of the whole world.

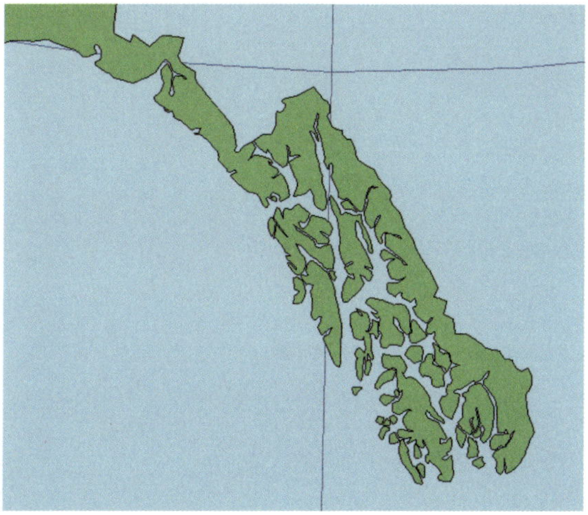

The State Plane Coordinate System uses Hotine Azimuth Natural Origin for the Alaskan panhandle.

DESCRIPTION

Also known as Oblique Cylindrical Orthomorphic.

This is an oblique rotation of the Mercator projection. Developed for conformal mapping of areas that are obliquely oriented and do not follow a north–south or east–west trend.

PROJECTION METHOD

Cylindrical. Oblique aspect of the Mercator projection. Oblique Mercator has several different types. You can define the tilt of the projection by either specifying two points or a point and an angle measuring east of north (the azimuth).

By default, the coordinate origin of the projected coordinates is located where the central line of the projection crosses the equator. As an example, if you use an Oblique Mercator (natural origin) for West Virginia, while the center of the projection is -80.75, 38.5, the natural origin is approximately -112.8253, 0.0. You can move the projection origin to the center of your data by using the Two-Point Center or Azimuth Center cases.

LINE OF TANGENCY

A single oblique great-circle line or secancy along two oblique small circles parallel to and equidistant from the central great circle.

LINEAR GRATICULES

Two meridians 180 degrees apart.

PROPERTIES

Shape

Conformal. Local shapes are true.

Area

Distortion increases with distance from the central line.

Direction

Local angles are correct.

Distance

True along the chosen central line.

LIMITATIONS

Use should be limited to regions near the central line. When using an ellipsoid, constant scale along the central line and perfect conformality cannot be maintained simultaneously.

USES AND APPLICATIONS

Ideal for conformal mapping of regions that have an oblique orientation.

Used for large-scale mapping in the Alaskan panhandle. Switzerland uses a different implementation of Oblique Mercator by Rosenmund, while Madagascar uses the Laborde version. These implementations aren't compatible.

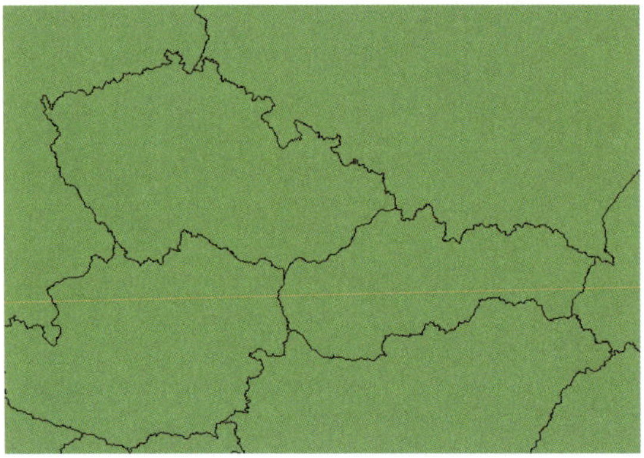

This example of the Krovak projection uses a right-handed coordinate system.

DESCRIPTION

This projection is an oblique case of the Lambert conformal conic projection and was designed in 1922 by Josef Krovak. Used in the Czech Republic and Slovakia. Also known as S-JTSK.

PROJECTION METHOD

Conic projection based on one standard parallel. An azimuth parameter tilts the apex of the cone from the North Pole to create a new coordinate system. A standard parallel in the new system, called a pseudostandard parallel, defines the shape of the cone. A scale factor is applied to the pseudostandard parallel to create a secant case.

LINES OF CONTACT

Two pseudostandard parallels.

LINEAR GRATICULES

None.

PROPERTIES

Shape

Small shapes are maintained.

Area

Minimal distortion within the boundaries of the countries.

Direction

Local angles are accurate throughout because of conformality.

Distance

Minimal distortion within the boundaries of the countries.

LIMITATIONS

Designed strictly for Czech Republic and Slovakia.

USES AND APPLICATIONS

Used for topographic and other mapping in Czech Republic and Slovakia. The coordinates are usually positive to the south and west.

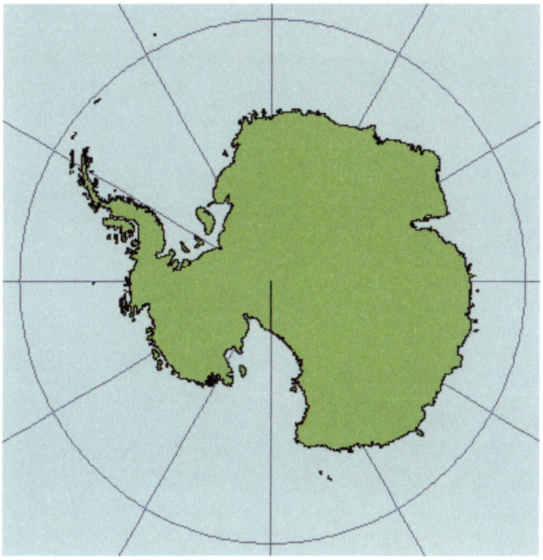

The central meridian is 0°, and the latitude of origin is 90° S.

DESCRIPTION

This projection preserves the area of individual polygons while simultaneously maintaining a true sense of direction from the center. The general pattern of distortion is radial. It is best suited for individual land masses that are symmetrically proportioned, either round or square.

PROJECTION METHOD

Planar, projected from any point on the globe. This projection can accommodate all aspects: equatorial, polar, and oblique.

POINT OF TANGENCY

A single point, located anywhere, specified by longitude and latitude.

LINEAR GRATICULES

All aspects—The central meridian defining the point of tangency.

Equatorial aspect—The equator.

Polar aspect—All meridians.

PROPERTIES

Shape

Shape is minimally distorted, less than 2 percent, within 15 degrees from the focal point. Beyond that, angular distortion is more significant; small shapes are compressed radially from the center and elongated perpendicularly.

Area

Equal area.

Direction

True direction radiating from the central point.

Distance

True at center. Scale decreases with distance from the center along the radii and increases from the center perpendicularly to the radii.

LIMITATIONS

The data must be less than a hemisphere in extent. The software cannot process any area more than 90 degrees from the central point.

USES AND APPLICATIONS

Population density (area).

Political boundaries (area).

Oceanic mapping for energy, minerals, geology, and tectonics (direction).

This projection can handle large areas; thus it is used for displaying entire continents and polar regions.

Equatorial aspect	Africa, Southeast Asia, Australia, the Caribbeans, and Central America
Oblique aspect	North America, Europe, and Asia

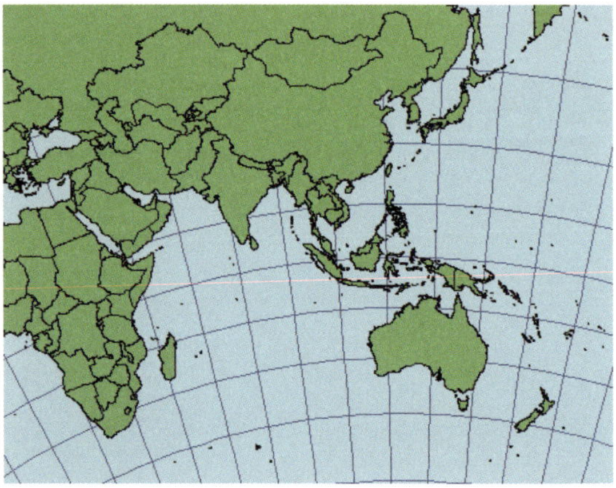

The central meridian is 125° E. The first and second standard parallels are 32° S and 7° N, while the latitude of origin is 32° S.

DESCRIPTION

This projection is one of the best for middle latitudes. It is similar to the Albers Conic Equal Area projection except that Lambert Conformal Conic portrays shape more accurately than area. The State Plane Coordinate System uses this projection for all zones that are more east–west in extent.

PROJECTION METHOD

Conic projection normally based on two standard parallels, making it a secant projection. The latitude spacing increases beyond the standard parallels. This is the only common conic projection that represents the poles as a single point.

LINES OF CONTACT

The two standard parallels.

LINEAR GRATICULES

All meridians.

PROPERTIES

Shape

All graticular intersections are 90 degrees. Small shapes are maintained.

Area

Minimal distortion near the standard parallels. Areal scale is reduced between standard parallels and increased beyond them.

Direction

Local angles are accurate throughout because of conformality.

Distance

Correct scale along the standard parallels. The scale is reduced between the parallels and increased beyond them.

LIMITATIONS

Best for regions predominantly east–west in extent and located in the middle north or south latitudes. Total latitude range should not exceed 35 degrees.

USES AND APPLICATIONS

SPCS for all east–west zones.

USGS 7½-minute quad sheets to match the State Plane Coordinate System.

Used for many new USGS maps created after 1957. It replaced the Polyconic projection.

Continental United States: standard parallels, 33° and 45° N.

Entire United States: standard parallels, 37° and 65° N.

DESCRIPTION

This is a specialized map projection that does not take into account the curvature of the earth. It's designed for very large-scale mapping applications. This projection is actually the Orthographic projection based on a spheroid.

PROJECTION METHOD

The coordinates of the center of the area of interest define the origin of the local coordinate system. The plane is tangent to the spheroid at that point, and the differences in z values are negligible between corresponding points on the spheroid and the plane. Because the differences in z values are ignored, distortions will greatly increase beyond roughly one degree from the origin.

USES AND APPLICATIONS

Large-scale mapping. In ArcInfo Workstation it should not be used for areas greater than one degree from the origin.

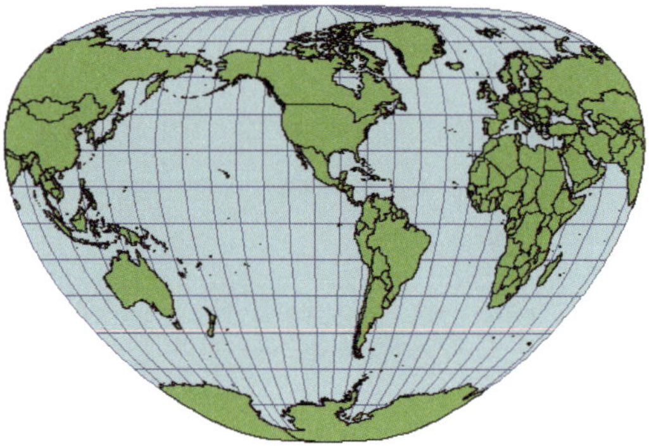

The central meridian is 100° W. The central parallel is 60° S.

DESCRIPTION

Karl Siemon created this pseudocylindrical projection in 1935. This projection was also presented in 1966 by Waldo Tobler. Loxodromes, or rhumb lines, are shown as straight lines with the correct azimuth and scale from the intersection of the central meridian and the central parallel.

PROJECTION METHOD

Pseudocylindrical. All parallels are straight lines, and all meridians are equally spaced arcs except the central meridian, which is a straight line. The poles are points.

LINEAR GRATICULES

The parallels and central meridian.

PROPERTIES

Shape

Shape is generally distorted. As the value of the central parallel increases from the equator, the overall shape of the world becomes more distorted.

Area

Generally distorted.

Direction

Directions are true only at the intersection of the central meridian and central latitude. Direction is distorted elsewhere.

Distance

Scale is true along the central meridian. It is constant along any latitude. The opposite latitude has a different scale if the central parallel isn't the equator.

LIMITATIONS

Useful only to show loxodromes.

USES AND APPLICATIONS

Suitable for displaying loxodromes.

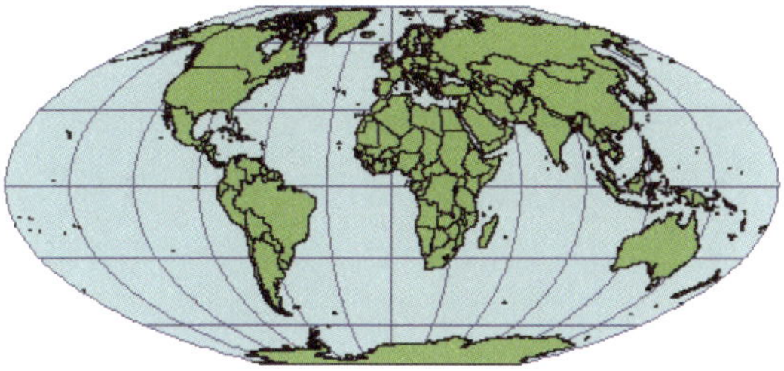

The central meridian is 0°.

DESCRIPTION

This equal-area projection is primarily used for world maps.

PROJECTION METHOD

A pseudocylindrical equal-area projection in which all parallels are straight lines and all meridians, except the straight central meridian, are equally spaced, fourth-order (quartic) curves.

LINEAR GRATICULES

All parallels are unequally spaced straight lines that are closer together at the poles. The poles are straight lines one-third as long as the equator. The central meridian is a straight line 0.45 times as long as the equator.

PROPERTIES

Shape

Shapes are stretched north–south along the equator, relative to the east–west dimension. This stretching decreases to zero at 33°45' N and S at the central meridian. Nearer the poles, features are compressed in the north–south direction.

Area

Equal area.

Direction

Distorted except at the intersection of 33°45' N and S and the central meridian.

Distance

Scale is distorted everywhere except along 33°45' N and S.

LIMITATIONS

Useful only as a world map.

USES AND APPLICATIONS

Thematic maps of the world.

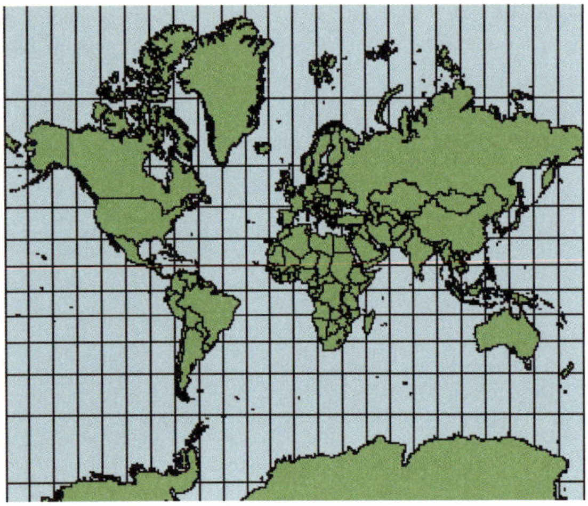

The central meridian is 0°.

DESCRIPTION

Originally created to display accurate compass bearings for sea travel. An additional feature of this projection is that all local shapes are accurate and clearly defined.

PROJECTION METHOD

Cylindrical projection. Meridians are parallel to each other and equally spaced. The lines of latitude are also parallel but become farther apart toward the poles. The poles cannot be shown.

LINES OF CONTACT

The equator or two latitudes symmetrical around the equator.

LINEAR GRATICULES

All meridians and all parallels.

PROPERTIES

Shape

Conformal. Small shapes are well represented because this projection maintains the local angular relationships.

Area

Increasingly distorted toward the polar regions. For example, in the Mercator projection, although Greenland is only one-eighth the size of South America, Greenland appears to be larger.

Direction

Any straight line drawn on this projection represents an actual compass bearing. These true direction lines are rhumb lines and generally do not describe the shortest distance between points.

Distance

Scale is true along the equator or along the secant latitudes.

LIMITATIONS

The poles cannot be represented on the Mercator projection. All meridians can be projected, but the upper and lower limits of latitude are approximately 80° N and S. Large area distortion makes the Mercator projection unsuitable for general geographic world maps.

USES AND APPLICATIONS

Standard sea navigation charts (direction).

Other directional uses: air travel, wind direction, ocean currents.

Conformal world maps.

The best use of this projection's conformal properties applies to regions near the equator, such as Indonesia and parts of the Pacific Ocean.

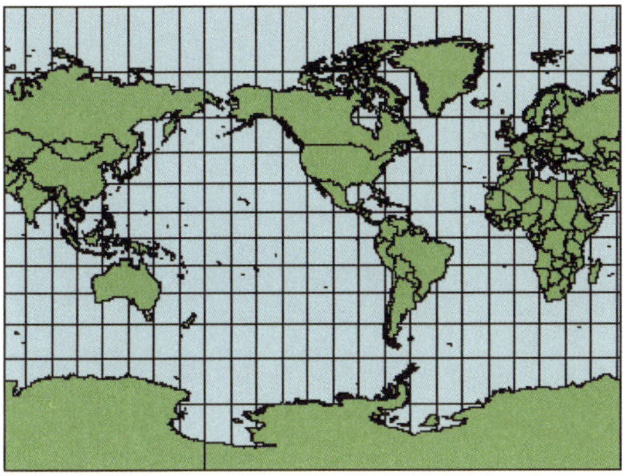

The central meridian is 118° W.

DESCRIPTION

This projection is similar to the Mercator projection except that the polar regions are not as areally distorted. Spacing between lines of latitude as they approach the poles is less than in the Mercator projection. It decreases the distortion in area, but the compromise introduces distortion in local shape and direction.

PROJECTION METHOD

Cylindrical projection. Meridians are parallel and equally spaced, lines of latitude are parallel, and the distance between them increases toward the poles. Both poles are represented as straight lines.

LINE OF CONTACT

The equator.

LINEAR GRATICULES

All meridians and all parallels.

PROPERTIES

Shape

Minimally distorted between 45th parallels, increasingly toward the poles. Land masses are stretched more east–west than they are north–south.

Area

Distortion increases from the equator toward the poles.

Direction

Local angles are correct only along the equator.

Distance

Correct distance is measured along the equator.

LIMITATIONS

Useful only as a world map.

USES AND APPLICATIONS

General-purpose world maps.

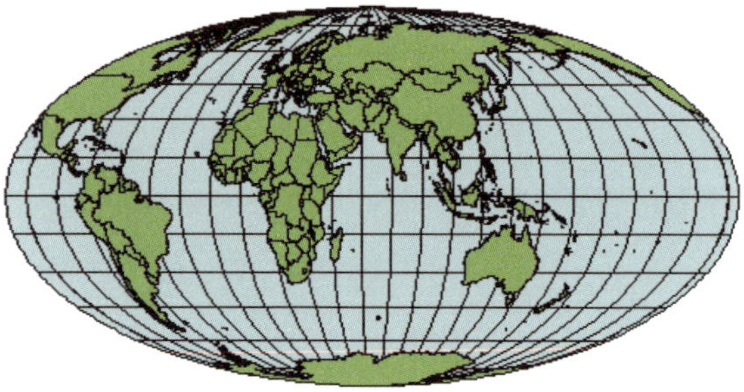

The central meridian is 65° E.

DESCRIPTION

Also called Babinet, Elliptical, Homolographic, or Homalographic.

Carl B. Mollweide created this pseudocylindrical projection in 1805. It is an equal-area projection designed for small-scale maps.

PROJECTION METHOD

Pseudocylindrical equal-area projection. All parallels are straight lines, and all meridians are equally spaced elliptical arcs. The exception is the central meridian, which is a straight line. The poles are points.

LINEAR GRATICULES

The equator and central meridian.

PROPERTIES

Shape

Shape is not distorted at the intersection of the central meridian and latitudes 40°44' N and S. Distortion increases outward from these points and becomes severe at the edges of the projection.

Area

Equal area.

Direction

Local angles are true only at the intersection of the central meridian and latitudes 40°44' N and S. Direction is distorted elsewhere.

Distance

Scale is true along latitudes 40°44' N and S. Distortion increases with distance from these lines and becomes severe at the edges of the projection.

LIMITATIONS

Useful only as a world map.

USES AND APPLICATIONS

Suitable for thematic or distribution mapping of the entire world, frequently in interrupted form.

Combined with the Sinusoidal to create Goode's Homolosine and Boggs.

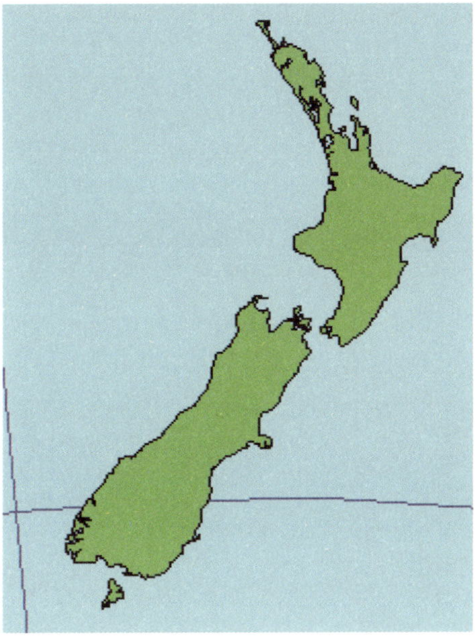

The central meridian is 173° E, and the latitude of origin is 41° S. The false easting is 2,510,000 meters, and the false northing is 6,023,150 meters.

DESCRIPTION

This is the standard projection for large-scale maps of New Zealand.

PROJECTION METHOD

Modified cylindrical. A sixth-order conformal modification of the Mercator projection using the International spheroid.

POINT OF TANGENCY

173° E, 41° S.

LINEAR GRATICULES

None.

PROPERTIES

Shape

Conformal. Local shapes are correct.

Area

Distortion is less than 0.04 percent for New Zealand.

Direction

Minimal distortion within New Zealand.

Distance

Scale is within 0.02 percent of true scale for New Zealand.

LIMITATIONS

Not useful for areas outside New Zealand.

USES AND APPLICATIONS

Used for large-scale maps of New Zealand.

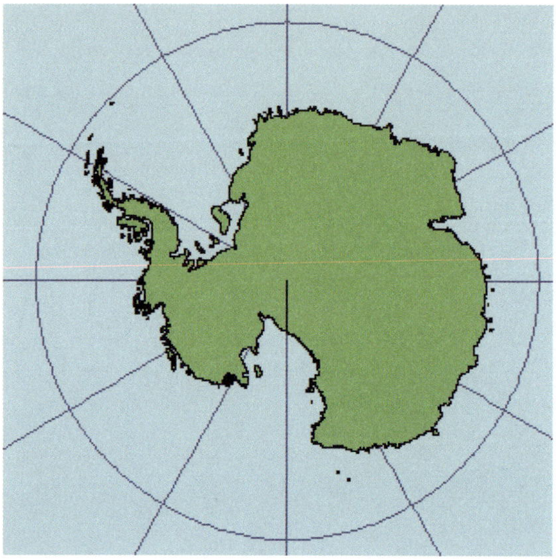

Central meridian is 0°, and latitude of origin is 90° S.

DESCRIPTION

This perspective projection views the globe from an infinite distance. This gives the illusion of a three-dimensional globe. Distortion in size and area near the projection limit appears more realistic to our eye than almost any other projection, except the Vertical Near-Side Perspective.

PROJECTION METHOD

Planar perspective projection, viewed from infinity. On the polar aspect, meridians are straight lines radiating from the center, and the lines of latitude are projected as concentric circles that become closer toward the edge of the globe. Only one hemisphere can be shown without overlapping.

POINT OF CONTACT

A single point located anywhere on the globe.

LINEAR GRATICULES

All aspects—The central meridian of the projection.

Equatorial aspect—All lines of latitude.

Polar aspect—All meridians.

PROPERTIES

Shape

Minimal distortion near the center; maximal distortion near the edge.

Area

The areal scale decreases with distance from the center. Areal scale is zero at the edge of the hemisphere.

Direction

True direction from the central point.

Distance

The radial scale decreases with distance from the center and becomes zero on the edges. The scale perpendicular to the radii, along the parallels of the polar aspect, is accurate.

LIMITATIONS

Limited to a view 90 degrees from the central point, a global hemisphere.

USES AND APPLICATIONS

Uses of this projection are aesthetic more than technical. The most commonly used aspect for this purpose is the oblique.

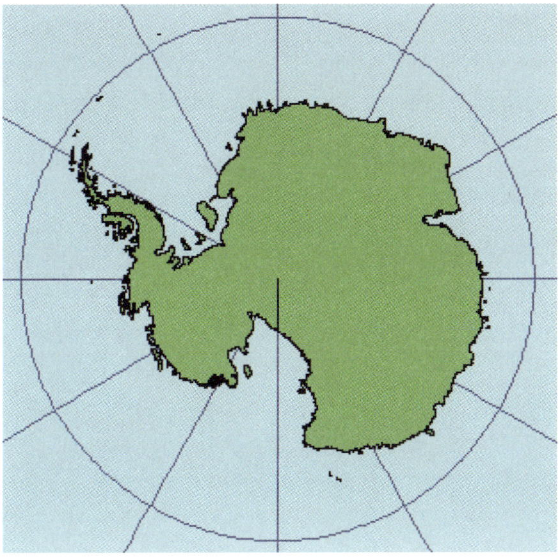

The central meridian is 0°, and the latitude of origin is 90° S.

DESCRIPTION

Also known as Vertical Near-Side Perspective or Vertical Perspective.

This projection is similar to the Orthographic projection in that its perspective is from space. In this projection, the perspective point is not an infinite distance away; instead, you can specify the distance. The overall effect of this projection is that it looks like a photograph taken vertically from a satellite or space vehicle.

PROJECTION METHOD

Planar perspective projection. The distance above the earth is variable and must be specified before the projection can be calculated. The greater the distance, the more closely this projection resembles the Orthographic projection. All aspects are circular projections of an area less than a hemisphere.

POINT OF CONTACT

A single point anywhere on the globe.

LINEAR GRATICULES

All aspects—The central meridian of the projection.

Polar aspect—All meridians.

Equatorial aspect—The equator.

PROPERTIES

Shape

Minimally distorted near the center, increasing toward the edge.

Area

Minimally distorted near the center; the area scale then decreases to zero on the edge or horizon.

Direction

True directions from the point of tangency.

Distance

Radial scale decreases from true scale at the center to zero on the projection edge. The scale perpendicular to the radii decreases, but not as rapidly.

LIMITATIONS

The actual range depends on the distance from the globe. In all cases, the range is less than 90 degrees from the center.

USES AND APPLICATIONS

Used as an aesthetic presentation rather than for technical applications.

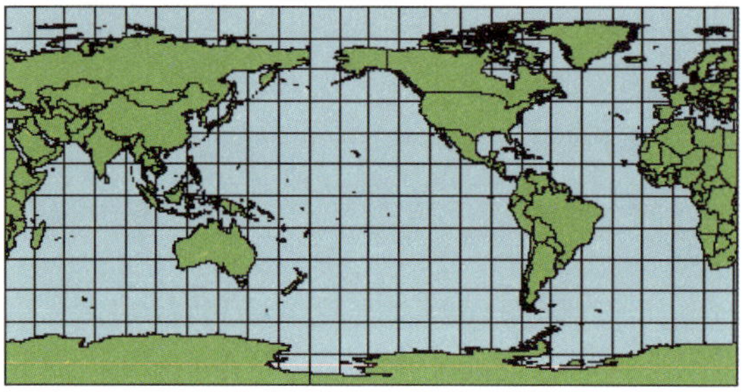

The central meridian is 149° W.

DESCRIPTION

Also known as Equirectangular, Equidistant Cylindrical, Simple Cylindrical, or Rectangular.

This projection is very simple to construct because it forms a grid of equal rectangles. Because of its simple calculations, its usage was more common in the past. In this projection, the polar regions are less distorted in scale and area than they are in the Mercator projection.

PROJECTION METHOD

This simple cylindrical projection converts the globe into a Cartesian grid. Each rectangular grid cell has the same size, shape, and area. All the graticular intersections are 90 degrees. The traditional Plate Carrée projection uses the equator as the standard parallel. The grid cells are perfect squares. In this projection, the poles are represented as straight lines across the top and bottom of the grid.

LINE OF CONTACT

Tangent at the equator.

LINEAR GRATICULES

All meridians and all parallels.

PROPERTIES

Shape

Distortion increases as the distance from the standard parallels increases.

Area

Distortion increases as the distance from the standard parallels increases.

Direction

North, south, east, and west directions are accurate. General directions are distorted, except locally along the standard parallels.

Distance

The scale is correct along the meridians and the standard parallels.

LIMITATIONS

Noticeable distortion of all properties away from standard parallels.

USES AND APPLICATIONS

Best used for city maps or other small areas with map scales large enough to reduce the obvious distortion.

Used for simple portrayals of the world or regions with minimal geographic data. This makes the projection useful for index maps.

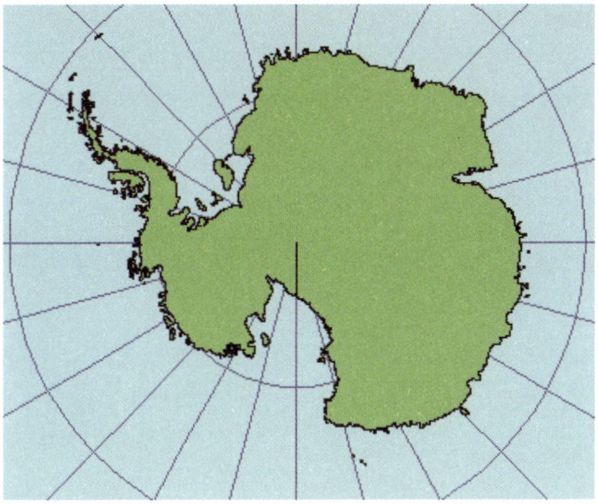

The central meridian is 0°, and the latitude of origin is 90° S.

DESCRIPTION

The projection is equivalent to the polar aspect of the Stereographic projection on a spheroid. The central point is either the North Pole or the South Pole. This is the only polar aspect planar projection that is conformal. The Polar Stereographic projection is used for all regions not included in the UTM coordinate system, regions north of 84° N and south of 80° S. Use UPS for these regions.

PROJECTION METHOD

Planar perspective projection, where one pole is viewed from the other pole. Lines of latitude are concentric circles. The distance between circles increases with distance from the central pole.

POINT OF TANGENCY

A single point, either the North Pole or the South Pole. If the plane is secant instead of tangent, the point of global contact is a line of latitude.

LINEAR GRATICULES

All meridians.

PROPERTIES

Shape

Conformal; accurate representation of local shapes.

Area

The farther from the pole, the greater the areal scale.

Direction

True direction from the pole. Local angles are true everywhere.

Distance

The scale increases with distance from the center. If a standard parallel is chosen rather than one of the poles, this latitude represents the true scale, and the scale nearer the pole is reduced.

LIMITATIONS

Normally not extended more than 90 degrees from the central pole because of increased scale and area distortion.

USES AND APPLICATIONS

Polar regions (conformal).

In the UPS system, the scale factor at the pole is 0.994, which corresponds to a latitude of true scale (standard parallel) at 81°06'52.3" N or S.

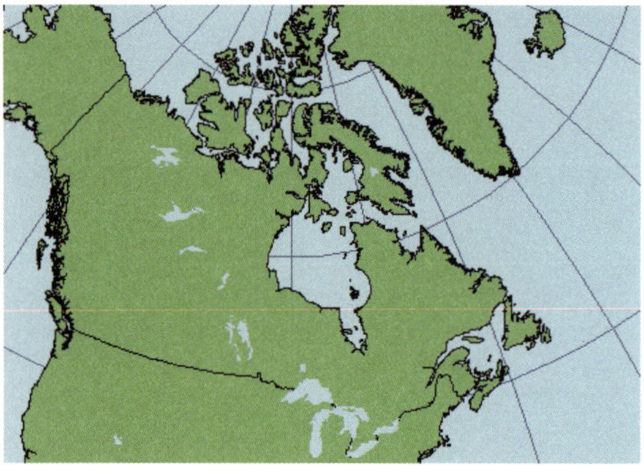

The central meridian is 90° W.

DESCRIPTION

The name of this projection translates into 'many cones'. This refers to the projection methodology. This affects the shape of the meridians. Unlike other conic projections, the meridians are curved rather than linear.

PROJECTION METHOD

More complex than the regular conic projections, but still a simple construction. This projection is created by lining up an infinite number of cones along the central meridian. This projection yields parallels that are not concentric. Each line of latitude represents the base of its tangential cone.

LINES OF CONTACT

Many lines; all parallels of latitude in the projection.

LINEAR GRATICULES

Central meridian of the projection and the equator.

PROPERTIES

Shape

No local shape distortion along the central meridian. Distortion increases with distance from the central meridian; thus, east–west distortion is greater than north–south distortion.

Area

Distortion in area increases with distance from the central meridian.

Direction

Local angles are accurate along the central meridian; otherwise, they are distorted.

Distance

The scale along each parallel and along the central meridian of the projection is accurate. Distortion increases along the meridians as the distance from the central meridian increases.

LIMITATIONS

Distortion is minimized on large-scale maps, such as topographic quadrangles, where meridians and parallels can be drawn in practice as straight-line segments. Producing a map library with this kind of map sheet is not advisable because errors accumulate and become visible when joining sheets in multiple directions.

USES AND APPLICATIONS

Used for 7½- and 15-minute topographic USGS quad sheets, from 1886 until approximately 1957. Note: Some new quad sheets after this date have been falsely documented as Polyconic. The present projection for east–west State Plane Coordinate System zones is Lambert Conformal Conic, and Transverse Mercator for north–south state zones.

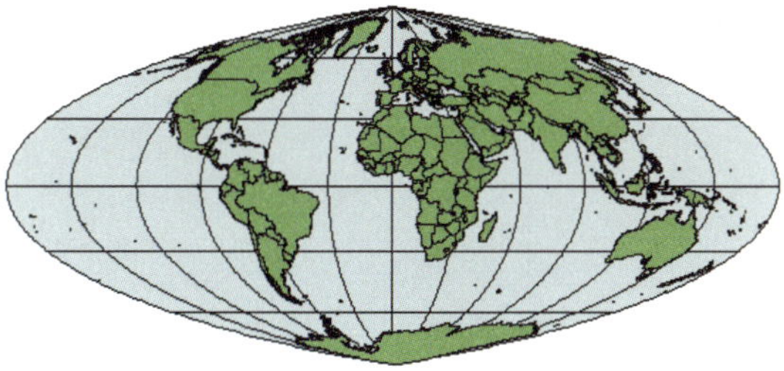

The central meridian is 0°.

DESCRIPTION

This pseudocylindrical equal-area projection is primarily used for thematic maps of the world.

PROJECTION METHOD

Pseudocylindrical equal-area projection.

LINEAR GRATICULES

The central meridian is a straight line 0.45 times the length of the equator. Meridians are equally spaced curves. Parallels are unequally spaced, straight parallel lines perpendicular to the central meridian. Their spacing decreases very gradually as they move away from the equator.

PROPERTIES

Shape

Generally distorted.

Area

Equal area.

Direction

Direction is generally distorted.

Distance

Scale is true along the equator. Scale is also constant along any given latitude and is symmetrical around the equator.

LIMITATIONS

Useful only as a world map.

USES AND APPLICATIONS

Thematic world maps. The McBryde–Thomas Flat-Polar Quartic projection is based on this projection.

The RSO projection is used for the island of Borneo.

DESCRIPTION

Also called RSO.

This projection is provided with two options for the national coordinate systems of Malaysia and Borneo and is similar to the Hotine Oblique Mercator projection.

PROJECTION METHOD

Oblique cylindrical projection. A line of true scale is drawn at an angle to the longitude of origin.

LINE OF CONTACT

A single, oblique, great-circle line.

LINEAR GRATICULES

Two meridians 180 degrees apart.

PROPERTIES

Shape

Conformal. Local shapes are true.

Area

Increases with distance from the center line.

Direction

Local angles are correct.

Distance

True along the chosen central line.

LIMITATIONS

Its use is limited to the areas along the central line. Although most of the world can be displayed, distortion increases rapidly away from the central line.

USES AND APPLICATIONS

Used for the national projections of Malaysia and Borneo.

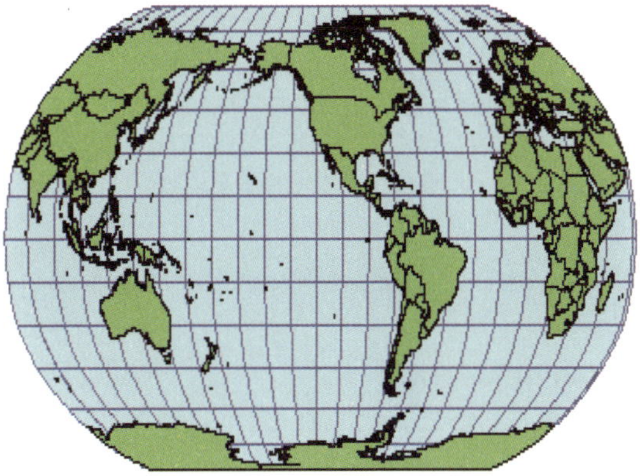

The central meridian is 118° W.

DESCRIPTION

Also called Orthophanic.

A compromise projection used for world maps.

PROJECTION METHOD

Pseudocylindrical. Meridians are equally spaced and resemble elliptical arcs, concave toward the central meridian. The central meridian is a straight line 0.51 times the length of the equator. Parallels are equally spaced straight lines between 38° N and S; spacing decreases beyond these limits. The poles are 0.53 times the length of the equator. The projection is based on tabular coordinates instead of mathematical formulas.

LINEAR GRATICULES

All parallels and the central meridian.

PROPERTIES

Shape

Shape distortion is very low within 45 degrees of the origin and along the equator.

Area

Distortion is very low within 45 degrees of the origin and along the equator.

Direction

Generally distorted.

Distance

Generally, scale is made true along latitudes 38° N and S. Scale is constant along any given latitude and for the latitude of the opposite sign.

LIMITATIONS

Neither conformal nor equal area. Useful only for world maps.

USES AND APPLICATIONS

Developed for use in general and thematic world maps.

Used by Rand McNally since the 1960s and by the National Geographic Society since 1988 for general and thematic world maps.

NOTE

The implementation in the desktop ArcGIS™ applications differs from the one in ArcInfo workstation. Use Robinson_ARC_INFO to access the ArcInfo workstation implementation.

The central meridian is 60° W. The first and second standard parallels are 5° S and 42° S. The latitude of origin is 32° S.

DESCRIPTION

Also called Equidistant Conic or Conic.

This conic projection can be based on one or two standard parallels. As the name implies, all circular parallels are an equal distance from each other, spaced evenly along the meridians. This is true whether one or two parallels are used.

PROJECTION METHOD

Cone is tangential if only one standard parallel is specified and secant if two standard parallels are specified. Graticules are evenly spaced. The space between each meridian is equal, as is the space between each of the concentric arcs that describe the lines of latitude. The poles are represented as arcs rather than points.

If the pole is given as the single standard parallel, the cone becomes a plane and the resulting projection is the same as a polar Azimuthal Equidistant.

If two standard parallels are placed symmetrically north and south of the equator, the resulting projection is the same as Equirectangular, and the Equirectangular projection must be used.

Use Equirectangular if the standard parallel is the equator.

LINES OF CONTACT

Depends on the number of standard parallels.

Tangential projections (Type 1)—One line, indicated by the standard parallel.

Secant projections (Type 2)—Two lines, specified as first and second standard parallels.

LINEAR GRATICULES

All meridians.

PROPERTIES

Shape

Local shapes are true along the standard parallels. Distortion is constant along any given parallel. Distortion increases with distance from the standard parallels.

Area

Distortion is constant along any given parallel. Distortion increases with distance from the standard parallels.

Direction

Locally true along the standard parallels.

Distance

True along the meridians and the standard parallels. Scale is constant along any given parallel but changes from parallel to parallel.

LIMITATIONS

Range in latitude should be limited to 30 degrees.

USES AND APPLICATIONS

Regional mapping of midlatitude areas that have a predominantly east–west extent.

Common for atlas maps of small countries.

Used by the former Soviet Union for mapping the entire country.

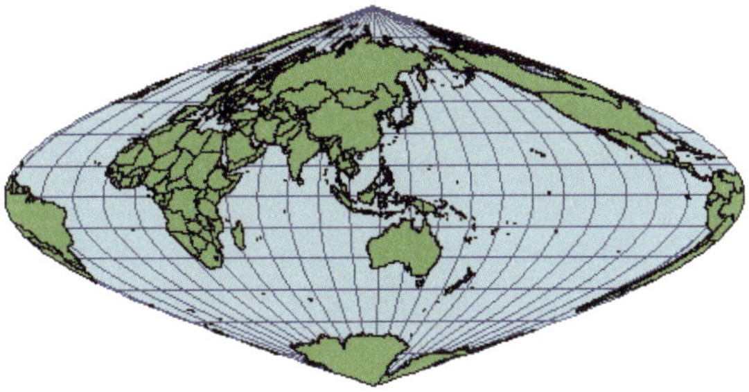

The central meridian is 117° E.

DESCRIPTION

Also known as Sanson–Flamsteed.

As a world map, this projection maintains equal area despite conformal distortion. Alternative formats reduce the distortion along outer meridians by interrupting the continuity of the projection over the oceans and by centering the continents around their own central meridians, or vice versa.

PROJECTION METHOD

A pseudocylindrical projection where all parallels and the central meridian are straight. The meridians are curves based on sine functions with the amplitudes increasing with the distance from the central meridian.

LINEAR GRATICULES

All lines of latitude and the central meridian.

PROPERTIES

Shape

No distortion along the central meridian and the equator. Smaller regions using the interrupted form exhibit less distortion than the uninterrupted sinusoidal projection of the world.

Area

Areas are represented accurately.

Direction

Local angles are correct along the central meridian and the equator but distorted elsewhere.

Distance

The scale along all parallels and the central meridian of the projection is accurate.

LIMITATIONS

Distortion is reduced when used for a single land mass rather than the entire globe. This is especially true for regions near the equator.

USES AND APPLICATIONS

Used for world maps illustrating area characteristics, especially if interrupted.

Used for continental maps of South America, Africa, and occasionally other land masses, where each has its own central meridian.

DESCRIPTION

This projection is nearly conformal and has little scale distortion within the sensing range of an orbiting mapping satellite such as Landsat. This is the first projection to incorporate the earth's rotation with respect to the orbiting satellite. For Landsat 1, 2, and 3, the path range is from 1 to 251. For Landsat 4 and 5, the path range is from 1 to 233.

PROJECTION METHOD

Modified cylindrical, for which the central line is curved and defined by the ground track of the orbit of the satellite.

LINE OF TANGENCY

Conceptual.

LINEAR GRATICULES

None.

PROPERTIES

Shape

Shape is correct within a few parts per million for the sensing range of the satellite.

Area

Varies by less than 0.02 percent for the sensing range of the satellite.

Direction

Minimal distortion within the sensing range.

Distance

Scale is true along the ground track and varies approximately 0.01 percent within the sensing range.

LIMITATIONS

Plots for adjacent paths do not match without transformation.

USES AND APPLICATIONS

Specifically designed to minimize distortion within the sensing range of a mapping satellite as it orbits the rotating earth.

Used to tie satellite imagery to a ground-based planar coordinate system and for continuous mapping of satellite imagery.

Standard format used for data from Landsat 4 and 5.

DESCRIPTION

Also known as SPCS, SPC, State Plane, and State.

The State Plane Coordinate System is not a projection. It is a coordinate system that divides the 50 states of the United States, Puerto Rico, and the U.S. Virgin Islands into more than 120 numbered sections, referred to as zones. Each zone has an assigned code number that defines the projection parameters for the region.

PROJECTION METHOD

Projection may be cylindrical or conic. See Lambert Conformal Conic, Transverse Mercator, and Hotine Oblique Mercator for methodology and properties.

WHY USE STATE PLANE

Governmental organizations and groups who work with them primarily use the State Plane Coordinate System. Most often, these are county or municipal databases. The advantage of using SPCS is that your data is in a common coordinate system with other databases covering the same area.

WHAT IS STATE PLANE

The State Plane Coordinate System was designed for large-scale mapping in the United States. It was developed in the 1930s by the U.S. Coast and Geodetic Survey to provide a common reference system to surveyors and mappers. The goal was to design a conformal mapping system for the country with a maximum scale distortion of one part in 10,000, then considered the limit of surveying accuracy.

Three conformal projections were chosen: the Lambert Conformal Conic for states that are longer east–west, such as Tennessee and Kentucky; the Transverse Mercator projection for states that are longer north–south, such as Illinois and Vermont; and the Oblique Mercator projection for the panhandle of Alaska, because it lays at an angle.

To maintain an accuracy of one part in 10,000, it was necessary to divide many states into zones. Each zone has its own central meridian or standard parallels to maintain the desired level of accuracy. The boundaries of these zones follow county boundaries. Smaller states such as Connecticut require only one zone, while Alaska is composed of 10 zones and uses all three projections.

This coordinate system is referred to here as the State Plane Coordinate System of 1927 (SPCS 27). It is based on a network of geodetic control points referred to as the North American Datum of 1927 (NAD 1927 or NAD27).

STATE PLANE AND THE NORTH AMERICAN DATUM

Technological advancements of the last 50 years have led to improvements in the measurement of distances, angles, and the earth's size and shape. This, combined with moving the origin of the datum from Meades Ranch in Kansas to the earth's center of mass for compatibility with satellite systems, made it necessary to redefine SPCS 27. The redefined and updated system is called the State Plane Coordinate System of 1983 (SPCS 83). The coordinates for points are different for SPCS 27 and SPCS 83. There are several reasons for this. For SPCS 83, all State Plane coordinates published by NGS are in metric units, the shape of the spheroid of the earth is slightly different, some states have changed the definition of their zones, and values of longitude and latitude are slightly changed.

Officially, SPCS zones are identified by their NGS code. When ESRI® implemented the NGS codes, they were part of a proposed Federal Information Processing Standard (FIPS). For that reason, ESRI identifies the NGS zones as FIPS zones. That proposed standard was withdrawn, but ESRI maintains the FIPS name for continuity.

Sometimes people use an older Bureau of Land Management (BLM) system. The BLM system is outdated and doesn't include codes for some of the new zones. The values also overlap. You should always use the NGS/FIPS codes.

The following zone changes were made from SPCS 27 to SPCS 83. The zone numbers listed below are FIPS zone numbers. In addition, false easting and northing, or origin, of most zones has changed.

California—California zone 7, SPCS 27 FIPS zone 0407, was eliminated and included in California zone 5, SPCS 83 FIPS zone 0405.

Montana—The three zones for Montana, SPCS 27 FIPS zones 2501, 2502, and 2503, were eliminated and replaced by a single zone, SPCS 83 FIPS zone 2500.

Nebraska—The two zones for Nebraska, SPCS 27 FIPS zones 2601 and 2602, were eliminated and replaced by a single zone, SPCS 83 FIPS zone 2600.

South Carolina—The two zones for South Carolina, SPCS 27 FIPS zones 3901 and 3902, were eliminated and replaced by a single zone, SPCS 83 FIPS zone 3900.

Puerto Rico and Virgin Islands—The two zones for Puerto Rico and the Virgin Islands, St. Thomas, St. John, and St. Croix, SPCS 27 FIPS zones 5201 and 5202, were eliminated and replaced by a single zone, SPCS 83 FIPS zone 5200.

UNIT OF LENGTH

The standard unit of measure for SPCS 27 is the U.S. Survey foot. For SPCS 83, the most common unit of measure is the meter. Those states that support both feet and meters have legislated which feet-to-meters conversion they use. The difference between the two is only two parts in one million, but that can become noticeable when datasets are stored in double precision. The U.S. Survey foot equals 1,200/3,937 m, or 0.3048006096 m.

EXAMPLES OF ZONE DEFINITIONS

Here are two examples of SPCS 83 parameters:

State	**Alabama East**	**Tennessee**
ZONE	**3101**	**5301**
FIPS Zone	**0101**	**4100**
Projection	**Transverse**	**Lambert**
Standard Parallels		**35°15'**
		36°25'
Central Meridian	**-85°50'**	**-86°00'**
Scale Factor Reduction at Central Meridian		
1:25,000	**1:15,000**	
Latitude of Origin	**30°30'**	**34°20'**
Longitude of Origin	**-85°50'**	**-86°00'**
False Easting	**200,000**	**600,000**
False Northing	**0**	**0**

USES AND APPLICATIONS

Used for standard USGS 7½- and 15-minute quad sheets.

Used for most federal, state, and local large-scale mapping projects in the United States.

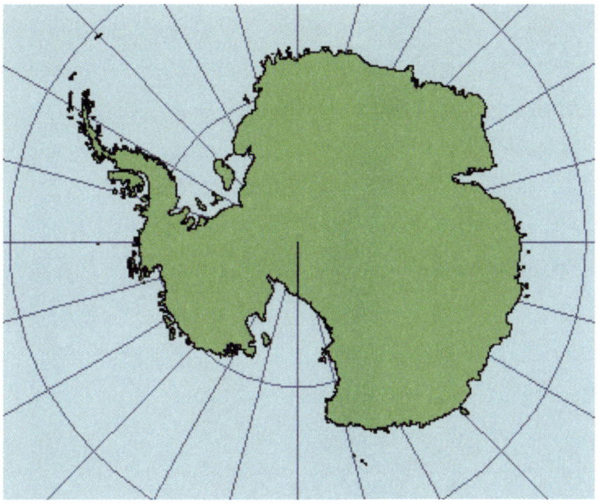

The central meridian is 0°, and the latitude of origin is 90° S.

DESCRIPTION

This projection is conformal.

PROJECTION METHOD

Planar perspective projection, viewed from the point on the globe opposite the point of tangency. Stereographic projects points on a spheroid directly to the plane. See Double Stereographic for a different implementation.

All meridians and parallels are shown as circular arcs or straight lines. Graticular intersections are 90 degrees. In the equatorial aspect, the parallels curve in opposite directions on either side of the equator. In the oblique case, only the parallel with the opposite sign to the central latitude is a straight line; other parallels are concave toward the poles on either side of the straight parallel.

POINT OF CONTACT

A single point anywhere on the globe.

LINEAR GRATICULES

Polar aspect—All meridians.

Equatorial aspect—The central meridian and the equator.

Oblique aspect—Central meridian and parallel of latitude with the opposite sign of the central latitude.

PROPERTIES

Shape

Conformal. Local shapes are accurate.

Area

True scale at center with distortion increasing with distance.

Direction

Directions are accurate from the center. Local angles are accurate everywhere.

Distance

Scale increases with distance from the center.

LIMITATIONS

Normally limited to one hemisphere. Portions of the outer hemisphere may be shown, but with rapidly increasing distortion.

USES AND APPLICATIONS

The oblique aspect has been used to map circular regions on the moon, Mars, and Mercury.

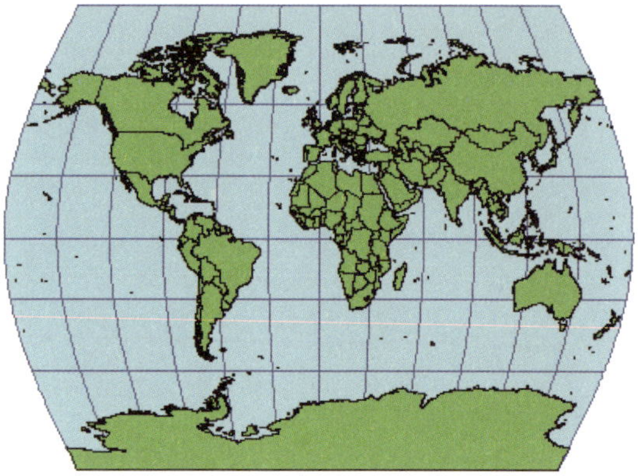

The central meridian is 0°.

DESCRIPTION

The Times projection was developed by Moir in 1965 for Bartholomew. It is a modified Gall's Stereographic, but Times has curved meridians.

PROJECTION METHOD

Pseudocylindrical. Meridians are equally spaced curves. Parallels are straight lines increasing in separation with distance from the equator.

LINES OF CONTACT

Two lines at 45° N and S.

LINEAR GRATICULES

All parallels and the central meridian.

PROPERTIES

Shape

Moderate distortion.

Area

Increasing distortion with distance from 45° N and S.

Direction

Generally distorted.

Distance

Scale is correct along parallels at 45° N and S.

LIMITATIONS

Useful only for world maps.

USES AND APPLICATIONS

Used for world maps by Bartholomew Ltd., a British mapmaking company, in *The Times Atlas*.

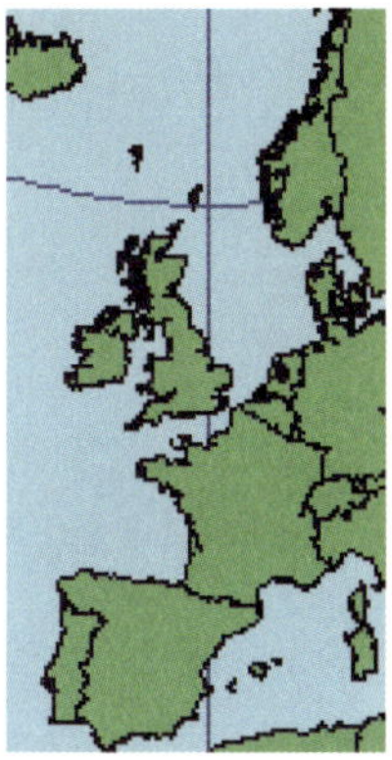

The central meridian and the latitude of origin are 0°. The scale factor is 1.0. Approximately 20 degrees of longitude are shown, which is close to the limit for Transverse Mercator.

DESCRIPTION

Also known as Gauss–Krüger (see that projection).

Similar to the Mercator except that the cylinder is longitudinal along a meridian instead of the equator. The result is a conformal projection that does not maintain true directions. The central meridian is placed in the center of the region of interest. This centering minimizes distortion of all properties in that region. This projection is best suited for north–south areas. The State Plane Coordinate System uses this projection for all zones that are more north–south in extent. The UTM and Gauss–Krüger coordinate systems are based on the Transverse Mercator projection.

PROJECTION METHOD

Cylindrical projection with central meridian placed in a particular region.

LINES OF CONTACT

Any single meridian for the tangent projection. For the secant projection, two almost parallel lines equidistant from the central meridian. For UTM, the lines are about 180 km from the central meridian.

LINEAR GRATICULES

The equator and the central meridian.

PROPERTIES

Shape

Conformal. Small shapes are maintained. Larger shapes are increasingly distorted away from the central meridian.

Area

Distortion increases with distance from the central meridian.

Direction

Local angles are accurate everywhere.

Distance

Accurate scale along the central meridian if the scale factor is 1.0. If it is less than 1.0, there are two straight lines with accurate scale equidistant from and on each side of the central meridian.

LIMITATIONS

Data on a spheroid or an ellipsoid cannot be projected beyond 90 degrees from the central meridian. In fact, the extent on a spheroid or ellipsoid should be limited to 15–20 degrees on both sides of the central meridian. Beyond that range, data projected to the Transverse Mercator projection may not project back to the same position. Data on a sphere does not have these limitations. In the desktop ArcGIS applications, you can use Transverse_Mercator_complex to access an implementation that does support data up to 80 degrees from the central meridian. Because of the complex mathematics involved, performance is affected.

USES AND APPLICATIONS

State Plane Coordinate System, used for predominantly north–south state zones.

USGS 7½-minute quad sheets. Most new USGS maps after 1957 use this projection, which replaced the Polyconic projection.

North America (USGS, central meridian scale factor is 0.926).

Topographic Maps of the Ordnance Survey of Great Britain after 1920.

UTM and Gauss–Krüger coordinate systems. The world is divided into 60 north and south zones six degrees wide.

Each zone has a scale factor of 0.9996 and a false easting of 500,000 meters. Zones south of the equator have a false northing of 10,000,000 meters to ensure that all y values are positive. Zone 1 is at 177° W.

The Gauss–Krüger coordinate system is very similar to the UTM coordinate system. Europe is divided into zones six degrees wide with the central meridian of zone 1 equal to 3° E. The parameters are the same as UTM except for the scale factor, which is equal to 1.000 rather than 0.9996. Some places also add the zone number times one million to the 500,000 false easting value. GK zone 5 could have false easting values of 500,000 or 5,500,000 meters.

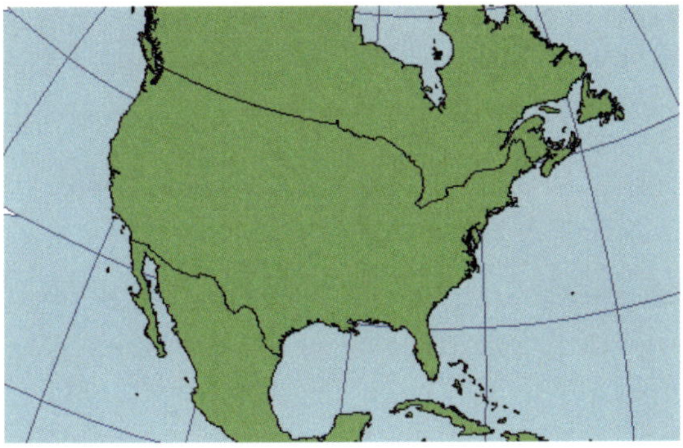

The first point is 117°30′ W, 34° N, and the second point is 83° W, 40° N.

DESCRIPTION

This projection shows the true distance from either of two chosen points to any other point on a map.

PROJECTION METHOD

Modified planar.

POINTS OF CONTACT

None.

LINEAR GRATICULES

Normally none.

PROPERTIES

Shape

Minimal distortion in the region of the two chosen points, if they're within 45 degrees of each other. Increasing distortion beyond this region.

Area

Minimal distortion in the region of the two chosen points, if they're within 45 degrees of each other. Increasing distortion beyond this region.

Direction

Varying distortion.

Distance

Correct from either of two chosen points to any other point on the map. Straight line from either point represents the correct great circle length but not the correct great circle path.

LIMITATIONS

Does not represent great circle paths.

USES AND APPLICATIONS

Used by the National Geographic Society for maps of Asia.

Adapted form used by Bell Telephone system for determining the distance used to calculate long distance telephone rates.

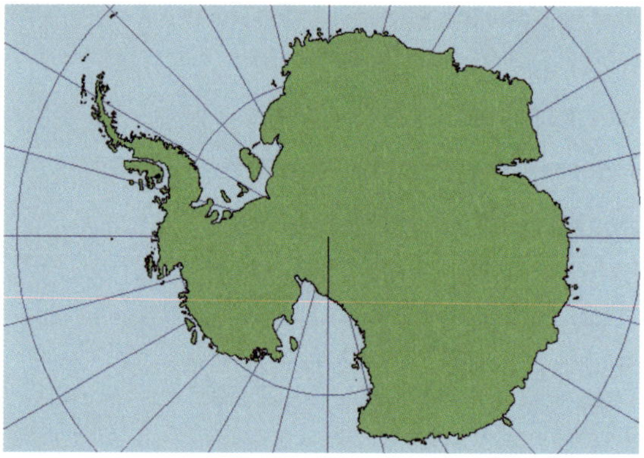

The central meridian is 90° S. The latitude of standard parallel is 81°06'52.3" S. The false easting and northing values are 2,000,000 meters.

DESCRIPTION

Also known as UPS.

This form of the Polar Stereographic projection maps areas north of 84° N and south of 80° S that aren't included in the UTM Coordinate System. The projection is equivalent to the polar aspect of the Stereographic projection of the spheroid with specific parameters. The central point is either the North Pole or the South Pole.

PROJECTION METHOD

Approximately (for the spheroid) planar perspective projection, where one pole is viewed from the other pole. Lines of latitude are concentric circles. The distance between circles increases away from the central pole. The origin at the intersection of meridians is assigned a false easting and false northing of 2,000,000 meters.

LINES OF CONTACT

The latitude of true scale, 81°06'52.3" N or S, corresponds to a scale factor of 0.994 at the pole.

LINEAR GRATICULES

All meridians.

PROPERTIES

Shape

Conformal. Accurate representation of local shape.

Area

The farther from the pole, the greater the area scale.

Direction

True direction from the pole. Local angles are correct everywhere.

Distance

In general, the scale increases with distance from the pole. Latitude 81°06'52.3" N or S has true scale. The scale closer to the pole is reduced.

LIMITATIONS

The UPS is normally limited to 84° N in the north polar aspect and 80° S in the south polar aspect.

USES AND APPLICATIONS

Conformal mapping of polar regions.

Used for mapping polar regions of the UTM coordinate system.

DESCRIPTION

Also known as UTM.

The Universal Transverse Mercator system is a specialized application of the Transverse Mercator projection. The globe is divided into 60 north and south zones, each spanning six degrees of longitude. Each zone has its own central meridian. Zones 1N and 1S start at -180° W. The limits of each zone are 84° N and 80° S, with the division between north and south zones occurring at the equator. The polar regions use the Universal Polar Stereographic coordinate system.

The origin for each zone is its central meridian and the equator. To eliminate negative coordinates, the coordinate system alters the coordinate values at the origin. The value given to the central meridian is the false easting, and the value assigned to the equator is the false northing. A false easting of 500,000 meters is applied. A north zone has a false northing of zero, while a south zone has a false northing of 10,000,000 meters.

PROJECTION METHOD

Cylindrical projection. See the Transverse Mercator projection for the methodology.

LINES OF CONTACT

Two lines parallel to and approximately 180 km to each side of the central meridian of the UTM zone.

LINEAR GRATICULES

The central meridian and the equator.

PROPERTIES

Shape

Conformal. Accurate representation of small shapes. Minimal distortion of larger shapes within the zone.

Area

Minimal distortion within each UTM zone.

Direction

Local angles are true.

Distance

Scale is constant along the central meridian but at a scale factor of 0.9996 to reduce lateral distortion within each zone. With this scale factor, lines lying 180 km east and west of and parallel to the central meridian have a scale factor of one.

LIMITATIONS

Designed for a scale error not exceeding 0.1 percent within each zone. Error and distortion increase for regions that span more than one UTM zone. UTM is not designed for areas that span more than a few zones.

Data on a spheroid or an ellipsoid cannot be projected beyond 90 degrees from the central meridian. In fact, the extent on a spheroid or ellipsoid should be limited to 15–20 degrees on both sides of the central meridian. Beyond that range, data projected to the Transverse Mercator projection may not project back to the same position. Data on a sphere does not have these limitations.

USES AND APPLICATION

Used for United States topographic quadrangles, 1:100,000 scale.

Many countries use local UTM zones based on the official geographic coordinate systems in use.

Large-scale topographic mapping of the former Soviet Union.

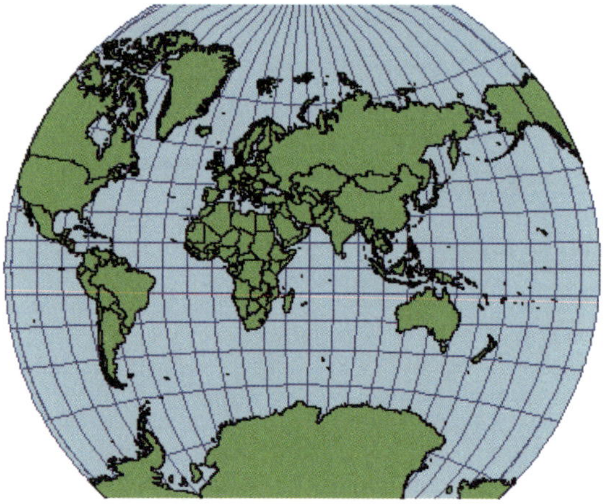

The central meridian is 56° E.

DESCRIPTION

This projection is similar to the Mercator projection except that it portrays the world with a curved graticule. The overall effect is that area is distorted less than on a Mercator projection, and the shape is distorted less than on equal-area projections.

PROJECTION METHOD

The Van der Grinten I projection is a compromise projection and is not in one of the more traditional classifications.

LINEAR GRATICULES

The equator and the central meridian of the projection.

PROPERTIES

Shape

Distortion increases from the equator to the poles.

Area

Minimal distortion along the equator and extreme distortion in the polar regions.

Direction

Local angles are correct only at the center.

Distance

Scale along the equator is correct.

LIMITATIONS

Can represent the world, but the most accurate representation is between the 75th parallels of latitude.

USES AND APPLICATIONS

Used for world maps.

Formerly the standard world map projection of the National Geographic Society.

Vertical Near-Side Perspective

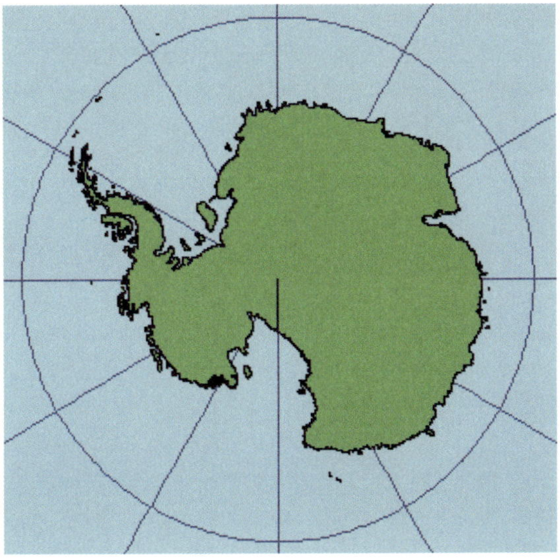

The central meridian is 0°, and the latitude of origin is 90° S.

DESCRIPTION

Unlike the Orthographic projection, this perspective projection views the globe from a finite distance. This perspective gives the overall effect of the view from a satellite.

PROJECTION METHOD

Planar perspective projection, viewed from a specified distance above the surface. All aspects are either circular or an area less than a full hemisphere.

Polar aspect—Meridians are straight lines radiating from the center, and the lines of latitude are projected as concentric circles that become closer toward the edge of the globe.

Equatorial aspect—The central meridian and the equator are straight lines. The other meridians and parallels are elliptical arcs.

POINT OF CONTACT

A single point located anywhere on the globe.

LINEAR GRATICULES

All aspects—The central meridian of the projection.

Equatorial aspect—The equator.

Polar aspect—All meridians.

PROPERTIES

Shape

Minimal distortion near the center; maximal distortion near the edge.

Area

Minimal distortion near the center; maximal distortion near the edge.

Direction

True direction from the central point.

Distance

The radial scale decreases with distance from the center.

LIMITATIONS

Limited to a view less than 90 degrees from the central point.

USES AND APPLICATIONS

Uses of this projection are aesthetic more than technical. The most commonly used aspect for this purpose is the oblique.

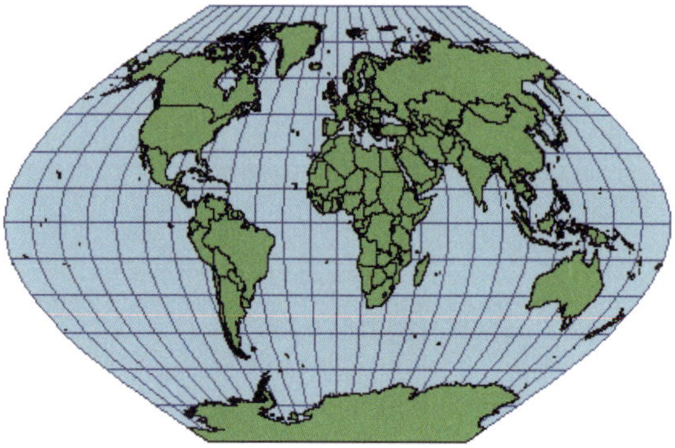

The central meridian is 0°.

DESCRIPTION

Often used for world maps, the Winkel I projection is a pseudocylindrical projection that averages the coordinates from the Equirectangular (Equidistant Cylindrical) and Sinusoidal projections. Developed by Oswald Winkel in 1914.

PROJECTION METHOD

Pseudocylindrical. Coordinates are the average of the Sinusoidal and Equirectangular projections. Meridians are equally spaced sinusoidal curves curving toward the central meridian. The central meridian is a straight line. Parallels are equally spaced straight lines. The length of the poles and the central meridian depends on the standard parallels. If the standard parallel is the equator, Eckert V results.

LINEAR GRATICULES

The parallels and the central meridian.

PROPERTIES

Shape

Generally distorted.

Area

Generally distorted.

Direction

Generally distorted.

Distance

Generally, scale is made true along latitudes 50°28' N and S.

LIMITATIONS

Neither conformal nor equal area. Useful only for world maps.

USES AND APPLICATIONS

Developed for use in general world maps. If the standard parallels are 50°28' N and S, the total area scale is correct, but local area scales vary.

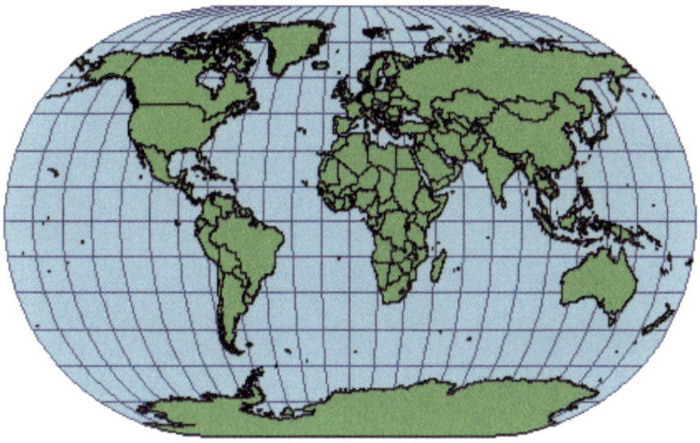

The central meridian is 0°.

DESCRIPTION

A pseudocylindrical projection that averages the coordinates from the Equirectangular and Mollweide projections. Developed by Oswald Winkel in 1918.

PROJECTION METHOD

Pseudocylindrical. Coordinates are the average of the Mollweide and Equirectangular projections. Meridians are equally spaced curves, curving toward the central meridian. The central meridian is a straight line. Parallels are equally spaced straight lines. The length of the poles and the central meridian depends on the standard parallels.

LINEAR GRATICULES

The parallels and the central meridian.

PROPERTIES

Shape

Generally distorted.

Area

Generally distorted.

Direction

Generally distorted.

Distance

Generally, scale is made true along the standard latitudes.

LIMITATIONS

Neither conformal nor equal area. Useful only for world maps.

USES AND APPLICATIONS

Developed for use in general world maps.

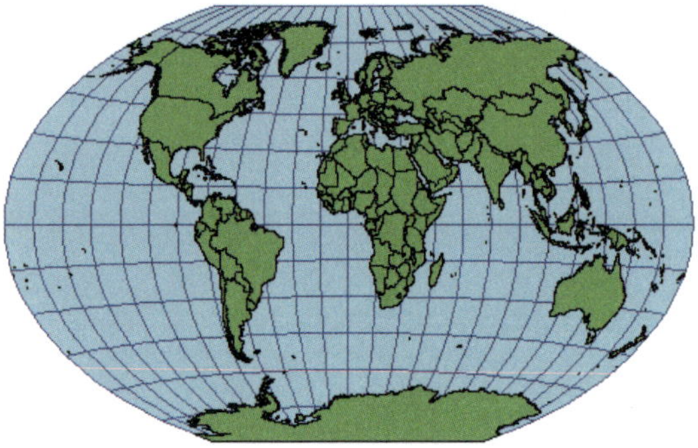

The central meridian is 0°, and the standard parallels are at 50.467° N and S.

DESCRIPTION

A compromise projection used for world maps that averages the coordinates from the Equirectangular (Equidistant Cylindrical) and Aitoff projections. Developed by Oswald Winkel in 1921.

PROJECTION METHOD

Modified azimuthal. Coordinates are the average of the Aitoff and Equirectangular projections. Meridians are equally spaced and concave toward the central meridian. The central meridian is a straight line. Parallels are equally spaced curves, concave toward the poles. The poles are around 0.4 times the length of the equator. The length of the poles depends on the standard parallel chosen.

LINEAR GRATICULES

The equator and the central meridian.

PROPERTIES

Shape

Shape distortion is moderate. In the polar regions along the outer meridians, the distortion is severe.

Area

Distortion is moderate. In the polar regions along the outer meridians, the distortion is severe.

Direction

Generally distorted.

Distance

Generally, scale is made true along latitudes 50.467° N and S or 40° N and S. The second case is used by Bartholomew Ltd., a British mapmaking company.

LIMITATIONS

Neither conformal nor equal area. Useful only for world maps.

USES AND APPLICATIONS

Developed for use in general and thematic world maps.

Used by the National Geographic Society since 1998 for general and thematic world maps.

Selected References

Datums, Ellipsoids, Grids and Grid Reference Systems. Washington, D.C.: NIMA, 1990. Technical Manual 8358.1, www.nima.mil/GandG/pubs.html.

Department of Defense World Geodetic System 1984. Third Edition. Washington, D.C.: NIMA, 1997. Technical Report 8350.2, www.nima.mil/GandG/pubs.html.

European Petroleum Survey Group, *EPSG Geodesy Parameters, v6.3.* www.epsg.org, 2003.

European Petroleum Survey Group, *POSC Literature Pertaining to Coordinate Conversions and Transformations including Formulas.* 2002. Guidance Note Number 7.

Geodesy for the Layman. Fourth Edition. Washington, D.C.: NIMA, 1984. Technical Report 80-003, www.nima.mil/GandG/pubs.html.

Hooijberg, Maarten, *Practical Geodesy: Using Computers.* Berlin: Springer–Verlag, 1997.

Junkins, D.R., and S.A. Farley, *NTv2 Developer's Guide.* Geodetic Survey Division, Natural Resources Canada, 1995.

Junkins, D.R., and S.A. Farley, *NTv2 User's Guide.* Geodetic Survey Division, Natural Resources Canada, 1995.

Maling, D.H., *Coordinate Systems and Map Projections.* Second Edition. Oxford: Pergamon Press, 1993.

National Geodetic Survey, NADCON Release Notes, README file accompanying NADCON Version 2.1. NOAA/NGS, July 2000.

Rapp, Richard H., *Geometric Geodesy: Part I.* Department of Geodetic Science and Surveying, Ohio State University, April 1991.

Rapp, Richard H., *Geometric Geodesy: Part II.* Department of Geodetic Science and Surveying, Ohio State University, March 1993.

Snyder, John P., *Map Projections: A Working Manual.* USGS Professional Paper 1395. Washington, D.C.: USGS, 1993.

Snyder, John P., and Philip M. Voxland, *An Album of Map Projections.* USGS Professional Paper 1453. Washington, D.C.: USGS, 1989.

Soler, T., and L.D. Hothem (1989), "Important Parameters Used in Geodetic Transformations." *Journal of Surveying Engineering* 112(4):414–417, November 1989.

Torge, Wolfgang, *Geodesy.* Second Edition. New York: de Gruyter, 1991.

Vanicek, Petr, and Edward J. Krakiwsky, *Geodesy: The Concepts.* Amsterdam: North-Holland Publishing Company, 1982.

Voser, Stefan A., *MapRef: The Collection of Map Projections and Reference Systems for Europe.* www.mapref.org, 2002.

Glossary

angular units

The unit of measurement on a sphere or a spheroid; usually degrees. Some map projection parameters, such as the central meridian and standard parallel, are defined in angular units.

aspect

The conceptual center of a projection system. See also equatorial, oblique, and polar aspect.

azimuth

The angle, measured in degrees, between a baseline drawn from a center point and another line drawn from the same point. Normally, the baseline points true north and the angle is measured clockwise from the baseline. Often used to define an oblique cylindrical map projection or the angle of a geodesic between two points.

azimuthal projection

A form of map projection in which the earth is projected onto a conceptual tangent or secant plane. Azimuthal projections are characterized by the fact that the azimuth, or direction, from the center of the projection to every other point on the map is shown correctly. Therefore the shortest route from the center point to any other point is shown as a straight line. See planar projection.

central meridian

The line of longitude that defines the center and often the x origin of a projected coordinate system. In planar rectangular coordinate systems of limited extent, such as state plane, grid north coincides with true north at the central meridian.

circle

A two-dimensional geometric shape for which the distance from the center to any point on the edge is equal; the closed plane curve defining such a shape or the surface bounded by such a curve.

conformal projection

A projection that preserves the correct shapes of small areas. Graticule lines intersect at 90-degree angles, and at any point on the map the scale is the same in all directions. A conformal projection maintains all angles at each point, including those between the intersections of arcs; therefore the size of areas enclosed by many arcs may be greatly distorted. No map projection can preserve the shapes of larger regions. Also called an orthomorphic projection (Snyder and Voxland, 1989).

conic projection

A projection made by projecting geographic features of the earth onto a tangent or secant cone that is wrapped around the globe in the manner of a party hat. The cone is then cut lengthwise and unrolled into a flat map.

coordinate system

A fixed reference framework superimposed onto the surface of an area to designate the position of a point within it; a reference system consisting of a set of points, lines and/or surfaces, and a set of rules, used to define the positions of points in space in either two or three dimensions. The Cartesian coordinate system and the geographic coordinate system used on the Earth's surface are common examples of coordinate systems.

cylindrical projection

A projection resulting from the conceptual projection of the earth's geographic features onto a tangent or secant cylinder wrapped around the globe. The cylinder is then cut and unrolled into a flat map (Snyder and Voxland, 1989).

datum

In the most general sense, any set of numeric or geometric constants from which other quantities, such as coordinate systems, can be defined. A datum defines a reference surface. There are many types of datums, but most fall into two categories: horizontal and vertical.

datum transformation

See geographic transformation.

eccentricity

A measure of how much an ellipse deviates from a circle, expressed as the ratio of the distance between the center and one focus of an ellipsoid to the length of its semimajor axis. The square of the eccentricity, e^2, is commonly used with the semimajor axis a to define a spheroid in map projection equations.

ellipse

A closed plane curve section of a right circular cone.

ellipsoid

1. A three-dimensional, closed geometric shape, all planar sections of which are ellipses or circles. An ellipsoid has three independent axes, and is usually specified by the lengths a,b,c of the three semi-axes. If an ellipsoid is made by rotating an ellipse about one of its axes, then two axes of the ellipsoid are the same, and it is called an ellipsoid of revolution, or spheroid. If the lengths of all three of its axes are the same, it is a sphere.
2. When used to represent the earth, an oblate ellipsoid of revolution, made by rotating an ellipse about its minor axis.

ellipticity

See eccentricity. See also flattening.

equal-area projection

A projection in which the whole of the map as well as each part has the same proportional area as the corresponding part of the earth. An equal area projection may distort shape, angle, scale, or any combination thereof (Snyder and Voxland, 1989). No flat map can be both equal area and conformal. Also known as an equivalent projection.

equator

The parallel of reference that is equidistant from the poles and defines the origin of latitude values as 0° north or south.

equatorial aspect

A planar—or azimuthal—projection with its central point located at the equator.

equidistant projection

A projection that maintains scale along one or more lines, or from one or two points to all other points on the map. Lines along which scale (distance) is correct are the same proportional length as the lines they reference on the globe. In the sinusoidal projections, for example, the central meridian and all parallels are their true lengths. An azimuthal equidistant projection centered on Chicago shows the correct distance between Chicago and any other point on the projection, but not between any other two points.

equivalent projection

See equal area projection.

false easting

The linear value added to all x-coordinates of a map projection so that none of the values in the geographic region being mapped are negative. See also false northing.

false northing

The linear value added to all y-coordinates of a map projection so that none of the values in the geographic region being mapped are negative. See false easting.

flattening

A measure of how much a spheroid differs from a sphere. The flattening is the ratio of the semimajor axis minus the semiminor axis to the semimajor axis.

Gauss–Krüger

A projected coordinate system that uses the Transverse Mercator projection to divide the world into standard zones six degrees wide. Used mainly in Europe and Asia, the Gauss-Krüger coordinate system is similar to the Universal Transverse Mercator (UTM) coordinate system.

geocentric coordinate system

A three-dimensional, earth-centered reference system in which locations are identified by their x, y, and z-values. The x axis is in the equatorial plane and intersects the prime meridian—usually Greenwich. The y axis is also in the equatorial plane and intersects the +90° meridian. The z axis coincides with the polar axis and is positive towards the north pole.

geocentric latitude

The angle between the equatorial plane and a line from a point on the surface to the center of the sphere or spheroid. On a sphere, all latitudes are geocentric. The unqualified term latitude generally refers to geographic, or geodetic, latitude.

geodesic

The shortest distance between two points on the surface of a spheroid. Any two points along a meridian form a geodesic.

geodetic datum

A datum designed to best fit all or part of the geoid, defined by the origin of an initial point—its latitude, longitude, and according to some authors, its elevation—the orientation of the network—the azimuth of a line from the origin—and two constants that define the reference spheroid. More recent definitions express the position and orientation of the datum as functions of the deviations in the meridian and in the prime vertical, the geoid-ellipsoid separation, and the parameters of a reference ellipsoid.

geodetic latitude

The angle that a line drawn perpendicular to the surface through a point on a spheroid makes with the equatorial plane.

geographic coordinate system

A reference system using latitude and longitude to define the locations of points on the surface of a sphere or spheroid.

geographic coordinates

A measurement of a location on the earth's surface expressed in degrees of latitude and longitude. See also geographic coordinate system.

geographic transformation

A method of converting data between two geographic coordinate systems (datums). Also known as a datum transformation.

global positioning system (GPS)

A constellation of 24 radio-emitting satellites deployed by the U.S. Department of Defense and used to determine location on the earth's surface. The orbiting satellites transmit signals that allow a GPS receiver anywhere on earth to calculate its own location through triangulation. The system is used in navigation, mapping, surveying, and other applications where precise positioning is necessary.

GPS

See global positioning system (GPS).

graticule

A network of longitude and latitude lines on a map or chart that relates points on a map to their true locations on the earth. (Snyder and Voxland, 1989).

great circle

Any circle or near circle produced by the intersection of a sphere and a flat plane that passes through the center of the sphere. The equator and all lines of longitude are great circles. Great circles are used in navigation, since the shortest path between any two points on the earth's surface lies on a great circle (Snyder and Voxland, 1989).

Greenwich prime meridian

The meridian adopted by international agreement in 1884 as the 0-degree meridian from which all other longitudes are calculated. The Greenwich prime meridian runs through the Royal Observatory in Greenwich, England.

grid

In cartography, any network of parallel and perpendicular lines superimposed on a map and used for reference. These grids are usually named after the map's projection; for example, Lambert grid and transverse Mercator grid.

HARN

See High Accuracy Reference Network (HARN).

High Accuracy Reference Network (HARN)

A resurvey of NAD 1983 control points using GPS techniques. The resurvey date is often included as part of the datum name—NAD 1983 (1991) or NAD91.

High Precision Geodetic (or GPS) Network

A resurvey of NAD 1983 control points using GPS techniques. The resurvey date is often included as part of the datum name—NAD 1983 (1991) or NAD91.

HPGN

See High Precision Geodetic (or GPS) Network.

international meridian

See Greenwich prime meridian.

interrupted projection
Discontinuities and gaps are added to a map to decrease the overall distortion. The world is divided, usually along certain meridians, into sections, or gores. Each section has its own projection origin.

latitude
The angular distance—usually measured in degrees—north or south of the equator. Lines of latitude are also called parallels. See geodetic latitude and geocentric latitude.

latitude of center
The latitude value that defines the center—and sometimes origin—of a projection.

latitude of origin
The latitude value that defines the origin of the y-coordinate values for a projection.

linear units
The unit of measurement on a plane or a projected coordinate system, often meters or feet. Map projection parameters, such as the false easting and false northing, are defined in linear units.

longitude
The angular distance—usually measured in degrees—east or west of a prime meridian.

longitude of center
The longitude value that defines the center—and sometimes origin—of a projection.

longitude of origin
The longitude value that defines the origin of the x-coordinate values for a projection.

loxodrome
See rhumb line.

major axis
The longer axis of an ellipse or spheroid.

map projection
See projection.

map scale
See scale.

meridian
A great circle on the earth that passes through the poles, often used synonymously with longitude. Meridians run north-south between the poles. From a prime meridian or 0 degrees longitude—usually the Greenwich prime meridian—measures of longitude are negative to the west and positive to the east, where they meet halfway around the globe at the line of 180 degrees longitude.

minor axis
The shorter axis of an ellipse or spheroid.

NAD 1927
North American Datum of 1927; also known as NAD27. The primary local geodetic datum and geographic coordinate system used to map the United States during the middle part of the 20th century, referenced to the Clarke spheroid of 1866 and an initial point at Meades Ranch, Kansas. Features on USGS topographic maps, including the corners of 7.5-minute quadrangle maps, are referenced to NAD27. It is gradually being replaced by the North American Datum of 1983. See also NAD 1983.

NAD 1983
North American Datum of 1983, also known as NAD83. A geocentric datum and graphic coordinate system based on the Geodetic Reference System 1980 ellipsoid (GRS80). Mainly used in North America, its measurements are obtained from both terrestrial and satellite data. See also NAD 1927.

oblate ellipsoid
An ellipsoid created by rotating an ellipse around its minor axis. The shape of the earth approximates an oblate ellipsoid with a flattening of one part in 298.257. See also ellipsoid, spheroid.

oblique aspect
See oblique projection.

oblate spheroid
See oblate ellipsoid.

oblique projection
1. A planar or cylindrical projection whose point of tangency is neither on the equator nor at a pole.
2. A conic projection whose axis does not line up with the polar axis of the globe.

3. A cylindrical projection whose lines of tangency or secancy follow neither the equator nor a meridian.

orthomorphic projection
See conformal projection.

parallel
An imaginary east–west line encircling the earth, parallel to the equator and connecting all points of equal latitude. Also, the representation of this line on a globe or map.

parameter
1. A variable that determines the outcome of a function or operation.
2. In a map projection, a group of parameters define a specific instance of a projection. Parameters differ for each projection and can include central meridian, standard parallel, scale factor, or latitude of origin.

planar coordinate system
A two-dimensional measurement system that locates features on a map based on their distance from an origin (0,0) along two axes, a horizontal x-axis representing east–west and a vertical y-axis representing north–south.

planar projection
A projection resulting from the conceptual projection of the earth onto a tangent or secant plane. Usually, a planar projection is the same as an azimuthal or zenithal projection, because it shows true direction (Snyder and Voxland, 1989).

polar aspect
A planar projection with its central point located at either the North or South Pole.

polar flattening
See flattening.

prime meridian
See Greenwich prime meridian.

projected coordinate system
A reference system used to locate x,y and z positions of point, line, and area features in two or three dimensions. A projected coordinate system is defined by a geographic coordinate system, a map projection, any parameters needed by the map projection, and a linear unit of measure.

projection
A method by which the curved surface of the earth is portrayed on a flat surface. This generally requires a systematic mathematical transformation of the earth's graticule of lines of longitude and latitude onto a plane. It can be visualized as a transparent globe with a light bulb at its center, casting lines of latitude and longitude onto a sheet of paper. Generally, the paper is either flat and placed tangent to the globe—a planar or azimuthal projection—or formed into a cone or cylinder and placed over the globe—cylindrical and conical projections. Every map projection distorts distance, area, shape, direction, or some combination thereof.

radius
The distance from the center to the outer edge of a circle or circular curve.

reference ellipsoid
See ellipsoid.

rhumb line
A complex curve on the earth's surface that crosses every meridian at the same oblique angle. A rhumb line path follows a single compass bearing; it is a straight line on a Mercator projection, or a logarithmic spiral on a polar projection. A rhumb line is not the shortest distance between two points on a sphere (Snyder and Voxland, 1989). See also great circle, sphere.

scale
The ratio or relationship between a distance or area on a map and on the corresponding distance or area on the ground, commonly expressed as a fraction or ratio. A map scale of 1/100,000 or 1:100,000 means that one unit of measure on the map equals 100,000 of the same unit on the earth.

scale factor
In a coordinate system, a value, usually less than one, that converts a tangent projection to a secant projection. Represented by 'k0' or 'k'. If a projected coordinate system doesn't have a scale factor, the standard lines of the projection have a scale of 1.0. Other points on the map have scales greater or less than 1.0. If a projected coordinate system has a scale factor, the defining parameters no longer have a scale of 1.0.

secant projection

A projection whose surface intersects the surface of the earth. A secant conic or cylindrical projection, for example, is recessed into the globe, intersecting it at two circles. At the lines of intersection the projection is free from distortion.

semimajor axis

The equatorial radius of a spheroid. Often known as 'a'.

semiminor axis

The polar radius of a spheroid. Often known as 'b'.

sphere

A three-dimensional shape whose center is equidistant from every point on its surface, made by revolving a circle around its diameter.

spherical coordinate system

A reference system using positions of latitude and longitude to define the locations of points on the surface of a sphere or spheroid.

spheroid

When representing the earth, the three-dimensional shape obtained by rotating an ellipse about its minor axis. This is an oblate ellipsoid of revolution, also called an ellipsoid.

standard line

A line on a sphere or spheroid that has no length compression or expansion after being projected. Commonly, a standard parallel or central meridian.

standard parallel

The line of latitude where the projection surface touches the surface. A tangent conic or cylindrical projection has one standard parallel, while a secant conic or cylindrical projection has two. A standard parallel has no distortion.

State Plane Coordinate System

A projected coordinate system used in the United States that divides each state into one or more zones to minimize distortion caused by the map projection. Also known as SPCS and SPC.

tangent projection

A projection whose surface touches the earth's without piercing it. A tangent planar projection touches the globe at one point, while tangent conic and cylindrical projections touch the globe along a line. At the point or line of tangency the projection is free from distortion.

true-direction projection

See azimuthal projection.

unit of measure

A standard quantity used for measurements, such as length, area, and height. See also angular units or linear units.

Universal Transverse Mercator

UTM.

UTM

Universal Transverse Mercator. A projected coordinate system that divides the world into 60 north and south zones, six degrees wide.

WGS 1984

World Geodetic System of 1984. A geocentric datum and geographic coordinate system created by the United States military. Also known as WGS84.

Index

T

U

V

W

X

Y